A WORLD WITHOUT HEART ATTACKS

HOW WE CAN BEAT HEART DISEASE – AND WHAT COMES NEXT

By Jeffrey L. Boone, MD
Boone Heart Institute
Denver, CO

With Contributions from Tyler Boone

First edition 2026

ISBN Paperback: 979-8-89576-223-3
ISBN Hardback: 979-8-89576-190-8

Published by:

AUTHORS on MISSION

Disclaimer

The content of this book is for informational purposes only and is not intended to diagnose, treat, cure, or prevent any condition or disease. This book is not intended as a substitute for consultation with a licensed practitioner. Please consult with your own physician or healthcare specialist regarding the suggestions and recommendations made in this book. The use of this book implies your acceptance of this disclaimer.

Endorsements

"Dr. Boone has worked to optimize my heart health for over two decades. I have benefitted from his help both personally and professionally. With Dr. Boone on your team, you know you are in good hands."

—John Elway
Hall of Fame Quarterback, Denver Broncos
2 Time Super Bowl Champion
Super Bowl XXXIII MVP

"Dr. Jeffrey Boone does more than treat heart disease. He listens deeply, explains clearly, and never settles for 'good enough.' This book reflects the same care and optimism he brings to his work with patients."

—Mike Shanahan
Former NFL Head Coach
2 Time Super Bowl Champion

"Dr. Boone is a rare physician who combines exceptional clinical excellence with a forward-looking embrace of modern and future technologies, instilling confidence through clarity, compassion, and integrity. This book reflects his visionary leadership and unwavering commitment to the human side of healing."

—Daniel W. Yohannes
U.S. Ambassador to the OECD (2014-2017)
Former CEO of the Millennium Challenge Corporation

"As one of the select few preventive cardiologists who were decades ahead of the field, Dr. Boone has long championed a precision-based approach to heart health. This book is a vital contribution from a true pioneer who helped shift our focus from managing risk factors to addressing the underlying disease."

—Amir Ahmadi, MD
Director, Inpatient Cardiology, Mount Sinai Fuster Heart at Morningside
Lead Scientific Advisor, Elucid

"Dr. Boone is uniquely driven to provide the best care and results; he is always one step ahead in his awareness of opportunities to improve outcomes. We wish there were more of him!"

—Richard Bard
Founder, Bard Capital
Ellis Island Medal of Honor Awardee from U.S. Congress

"I have had the honor and privilege of knowing and working with Dr. Jeff Boone for 35+ years. What a blessing he has been to the Denver Broncos Football Club and me personally! Dr. Boone is and has been way ahead of the Cardiovascular Field. His preventive and treatment philosophies are very progressive and scientifically backed. Most of all he is a Great Man with Great Communication Skills. Simply, Dr Boone saves lives!"

—Steve "Greek" Antonopulos
Retired Director of Sports Medicine and Head Athletic Trainer
Denver Broncos Football Club
NFL Hall Of Fame Award of Excellence 2024

"When I met Dr Boone, I had a number of red flags that were hurtling me toward real mid-life problems. Extremely elevated cholesterol, plaque forming, and more. In a short period of time, all of that has reversed. My blood picture could not look better. I am exercising vigorously —more than I did a decade ago. (I hiked the Grand Canyon rim to rim a month ago.) More importantly, I have a complete picture and full confidence in my health. Dr Boone is truly a gift. I cannot thank him enough."

—Gary Ross
Academy Award Nominated Director & Screenwriter

"In professional sports, you always look at new approaches and new technology to get an edge. Dr. Boone is continually upgrading his program and sharpening his skills like an athlete. When I was evaluated by Dr. Boone, the testing exposed serious problems. I know I have a purpose, and thanks to Dr. Boone, I know I'm in the best hands. He has saved my life."

—Steve Hess
Founder, Hess Elite Training
Denver Nuggets Strength & Conditioning Coach, 1996 - 2017

"Dr. Boone should be considered the Godfather of preventive cardiology. Dr. Boone has been on the cutting edge of heart health for athletes during their career and post career. Dr. Boone's actionable guidance helps everyone prolong their lifespan. I would recommend to anyone to listen when Dr. Boone has something to say!"

—Keith "Doogie" Dugger
Head Athletic Trainer - Colorado Rockies

Table of Contents

Introduction

The Greatest Killer

The greatest epidemic in the history of the world moves in silence. It does not make you feel particularly bad. It does not change the way you look. It comes with no fever or weakness. It gives almost no warning before it strikes. Heart disease grows throughout our lives at a glacial pace, gathering undetected within the millimeters-thick lining of our arteries until it is too late for even a Hail Mary attempt to fix it. The first symptom is often death.

Such a counterintuitive disease—widespread yet difficult to detect, deadly yet not uncomfortable—was always going to be difficult to treat. In the century since heart disease (and its cousin, stroke) became the world's leading cause of death, the entire medical profession has worked diligently to contain the epidemic, with some success. Thanks to decades of ongoing education, everyone now knows you should exercise, eat right, and quit smoking. The signs of heart attack or stroke—tightness in the chest, pain in the left arm, the smell of burning toast—are reported ubiquitously in our media. Our first responders (and even many civilians) have become excellent at recognizing and reacting to the onset of a cardiac event, treating with an aspirin or providing CPR, or utilizing one of the defibrillators that hang inconspicuously on the walls in many public places. Once the patient is transported to the hospital, our highly skilled cardiologists and cardiac surgeons have become masters of emergency bypass surgery and other lifesaving procedures. This massive and costly apparatus keeps many alive who would have died a generation earlier.

And yet the epidemic continues, nearly unabated.

In fact, heart disease is projected to worsen in the coming decades. Despite all the remarkable technology and skill in our well-meaning healthcare system, someone dies of heart disease every 1.7 seconds. Indeed, each year there are 10 percent more deaths from cardiovascular disease (CVD) than from all cancers combined. CVD is also thirteen times more deadly than breast cancer, killing one in three women compared to the one in thirty-nine women that breast cancer kills. And in 2020, at the height of the pandemic, there were four times more CVD deaths than deaths from COVID-19.

Heart disease became the leading cause of death in 1921, more than a century ago. And despite all our skill, innovation, and investment over the last hundred years, it isn't getting much better. It is still thought of, by both the public and the medical field, as a "silent killer" because of the way it grows undetected and strikes seemingly without warning. We need to change this paradigm. We've learned in recent years that heart disease has measurable, anatomical predictors that ought to be known and measured in every person over forty. It's not that CVD fails to warn us of impending doom; it's that we fail to listen. Rather than trying to get ahead of heart disease in the same way that we screen for cancer with mammograms and colonoscopies, the majority of our attention and investment comes in our reactions after CVD strikes, in the form of bypasses, stents, and other fixes. This disease, despite its counterintuitive nature, is not silent at all. In fact, now that heart disease is killing 700,000 people per year in the United States alone, I'd say heart disease has actually become quite loud. It's time for us all to start listening.

A Tale of Two Fathers

Two photos sit on my living room mantel in a place of honor. In the first picture, one of my favorites, my late father, Stewart sits in the middle of a couch. I am sitting next to him on his right, while Tyler, my oldest son and Stewart's grandson, sits to his left. Lying on my father's lap is his infant great-grandson, Miles (my grandson). My father was ninety-four at the time, and Miles was less than six months old. It still warms my heart to see four generations of Boone men, all oldest sons, sitting together for one perfect moment.

In his youth, my father served as a bugler and artilleryman in World War II, where he earned a bronze star for his actions in the Battle of the Bulge. During the war, the military provided a ration of four cigarettes a day, to help the soldiers calm their nerves. My father, never a smoker, traded his cigarettes for French fries. After the war, he returned to Kansas, enrolled in college on a GI Bill scholarship, and met my mother, Joyce. My parents had both been raised in that same Kansas landscape, and my mother's family had survived the Dust Bowl. She once told me that the only time she saw fear in her father's eyes was when a black dust cloud the size of the Rocky Mountains appeared on the horizon, whirling toward their home as family members scrambled to find shelter. She remembered the dust settling so thick around their house that she could walk up the drift directly onto the roof. Despite all this, their family stayed in Kansas, rather than moving on toward the West Coast with so many other migrants.

After my parents married, my father started a small business, had three kids, presided over the Rotary Club, and played his trumpet for anyone who would listen, including a packed stadium of over forty thousand people at the BOLDERBoulder Memorial Day celebration every year from 2010 to 2022, when he was in his mid-eighties and nineties. In the last years of his life, my sister asked him to rate the quality of each day from one to ten, and he never gave her a number below seven.

In my own life, I became a medical doctor and developed and refined the Boone Heart Protocol, an elite preventive cardiology program designed to prevent heart attacks, strokes, and dementia, and my father was our in-house longevity experiment. By the time he passed at age ninety-eight, he had been on the Boone Heart Protocol for years. He had the heart of a man in his sixties, stretching his possible health span well into the tenth decade of life. Our hearts beat around a hundred thousand times in a day; in the average lifetime, they will beat more than 2.5 billion times. At my father's age, his heart had clocked in at an impressive total of over 3.5 billion beats! We used to joke that when he died, we would have to beat his heart with a stick to get it to stop. To be fair, his heart had received some mechanical help from modern medicine when he was in his nineties: a heart valve replacement, a cardiac ablation to fix an irregular rhythm, and a blocked artery unclogged. It seemed that just when he needed something, modern medicine arrived with the right solution.

On top of that, his brain remained healthy. His last scan, several months before he passed, revealed that he had the brain of someone half his age. He'd just finished coauthoring his fourth children's book with my sister three weeks before he passed. Eventually, he was killed, like many of our ancestors before us, by a common infection. While his organs were strong, his body had worn down from many happy decades of wear and tear, and ultimately he succumbed peacefully after a short battle with pneumonia. In his absence, however, Miles and his other great-grandchildren will cherish many memories of "Super Grandpa Stew's" final decade.

The second photo on our mantel shows Don Felgate, my wife's father, fishing with my son Tyler, Miles's father, when Tyler was just two, in a little pond outside of Portland, where we were living during my medical residency.

Don was not as fortunate as my father. Originally from Watertown, South Dakota, Don joined the military police as World War II ended and helped bring stability to Europe. He was the strong, silent type—six feet tall, athletic, with a stoic personality. Like many young men serving at the time, he found the proffered cigarettes to be calming in the midst of that stressful situation, and brought his smoking habit home with him after the service. Who could blame him for seeking a little relief after witnessing the unspeakable atrocities of Nazi Germany and the liberation of the concentration camps? Upon his return to the U.S., he started working at Southwestern Bell Telephone, before it became part of AT&T, and met my wife's mother, Beth. They settled down in a nice suburban area in Sioux Falls, South Dakota, and had eight children. When my wife, Colette, was a senior in high school, Don was promoted and the family moved to Des Moines, Iowa where they stayed until he retired in 1982.

Don loved to fish. His sister-in-law, my wife's aunt, owned a beach resort on the north coast of Washington State on the Strait of Juan de Fuca. Don's plan was to move there and help her manage the resort while spending the rest of his days fishing in the Puget Sound.

Not long after the picture on the mantel was taken, my father-in-law was fishing for salmon with two of his sons in the early morning. As they stood in the boat on the Strait, Don suddenly collapsed. His boys tried to revive him, performing CPR, but it didn't help; he died there on that boat at age fifty-seven. Tyler was the only grandchild to have met his grandfather.

Decades later, we learned that Don had a gene called lipoprotein(a), a very sticky and clot-generating kind of cholesterol, which causes a quicker buildup of plaque in the arteries. This gene is elevated in about one out of five people, and is especially exacerbated by smoking. The heart attack that killed him was undoubtedly caused by this unlucky combination of genetics and a habit that few in his generation escaped from, at a time when there was still little known about the health risks of cigarettes. I saw firsthand the grief and devastation that this disease could wreak. While we enjoyed the company of my father for another four decades after Don's passing, Don's spirit reminded us of all we had lost from heart disease. Indeed, how many of us are haunted by the ghosts of our beloved fathers, mothers, sisters, brothers, and children, who died too early from this sickness? Don's death motivated me to help find the answers so that we all—my own father, I myself, and you, too—might know our grandchildren and great-grandchildren.

We all want to live long, healthy lives like my father. Unfortunately, today's average lifespan for women is about seventy-nine years, while for men it is just seventy-three years. With good reason, we tend to view stories like my father-in-law's as the norm and my father's as the exception. And yet, there's simply no reason for his story to be an anomaly. Heart disease is the number-one killer in America, and it is entirely preventable. We have all the knowledge, medical innovations, tests, and pharmaceuticals we need to completely eradicate this disease from the face of the planet, just as we have done with polio, smallpox, tuberculosis, and dozens of other deadly diseases. A massive research project completed in 2021 analyzed over half a million anonymized human blood samples for what scientists call the "biomarkers of aging." The result of this study suggested that the human lifespan might have a cap of somewhere between 120 and 150 years. Interestingly, this aligns with what's written in the Bible in Genesis 6:3, when God says that mankind's days shall be limited to 120 years. With some care, education, and proper maintenance, we might all reasonably expect to live to a healthy hundred years and beyond.

So the question then becomes: Why are we losing the war on heart disease? And how do we finally win?

My Journey to Preventive Cardiology

I didn't grow up wanting to go into medicine. I was going to play in the National Basketball Association (NBA). I had been a good athlete all my life, especially a good basketball player. Growing up in a town of fourteen thousand in the 1960s, there wasn't much to do, so I spent most of my early days shooting hoops until the sun went down. My love of sports led me to a physical education major in college. I figured that on my way to that NBA career, I might learn more about exercise, nutrition, and how the body works. And if the NBA didn't work out, I felt my calling might be as a recreation director and physical education teacher.

As I went out into the world, I soon learned that the NBA was not interested in a slow, six-foot-four shooting guard who couldn't jump. After I graduated from Southwestern College, I went on to explore my physical education interest at Kansas State University, where I received a student teaching fellowship as I worked toward a master's degree. It was at Kansas State that I discovered a whole new world of the scientific and academic possibilities afforded by physical exercise. The focus in exercise studies at this time was shifting from sports performance to overall health and longevity. A few professors there had pioneered a freshman college course that approached exercise and wellness like an academic venture. I was one of the graduate students who would teach that course to freshmen. Exploring biomechanics in exercise physiology and human performance from a more scientific and medical perspective was quite a revolutionary concept at the time. I completed my master's degree coursework in 1974 and looked for a way to practice what I had been learning.

After a brief search, I was offered a job in Des Moines as the fitness director of the local YMCA. My salary was $9,000 per year. In some ways, it was the best job I ever had.

In those days, the YMCA was often the hub of physical fitness in a city, as the concept of the modern health club or gym had not yet taken hold. Because of this, healthcare leaders had begun to look at YMCAs as unique touchpoints for urban populations. Specifically, a partnership emerged between the YMCA and the American Heart Association with the goal of developing cardiac rehabilitation programs at YMCAs where patients recovering from heart attacks could be supervised and rehabilitated in a safe, standardized way.

At the Des Moines YMCA, I was honored to be the exercise physiologist acting as the "boots on the ground" for our area's cardiac rehabilitation patients, who were referred by internists and cardiologists from the Des Moines hospitals for help in their post-heart attack recovery process. I went to training sessions, learned how to use new monitoring equipment, and hired all the necessary personnel to work with these heart attack victims. Our team was primed and motivated to expand our reach from fitness into healthcare, when we ran into a snag: we couldn't find anyone to be our medical director.

I had no idea why we needed a doctor on board to implement our program. After all, weren't doctors referring their patients to us in the first place? But I came to understand that legally, we could not implement our program without the oversight of a doctor. I reached out to the doctors who had begun referring their patients to our program, but they just wanted to use these new services. None of them had the time or the interest to actually supervise the program. We stalled for a few weeks until, finally, one doctor begrudgingly agreed to become our medical director. He really didn't have to do much; it was just the power of his name and his degree as an MD we needed to launch the program, which we did with great success.

In addition to fitness training, I was regularly counseling various members of the YMCA community about their health. In those days, heart medications weren't very good. I sometimes say that I'd rather die than take some of those early medications for blood pressure and various other cardiac abnormalities; the side effects were brutal, and I saw them firsthand as I worked with our patients. Naturally, I began thinking about how to help patients avoid the need for these drugs in the first place.

In those days, most physicians weren't very aggressive about counseling patients toward good health practices. It required a great deal of coaxing to motivate people to exercise, lose weight, quit smoking, manage stress, and eat a healthy diet. Most physicians were consumed with caring for their already sick patient population. There wasn't enough bandwidth for them to also educate their relatively healthy patients about the hidden killers growing inside them because of their poor lifestyle choices. As a fitness counselor, I became acutely aware of this void in healthcare. As you'll see later, not much has changed.

At the same time, I saw the power and authority that having a medical degree afforded, even if that person was simply a figurehead. As obvious as it may seem, though, I still didn't make the leap to pursuing a career in medicine right away. Instead, one day, while floating in a pool in Lake Tahoe, I was talking about the future with my soon-to-be wife, Colette. I told her I was thinking of going back to school to get my PhD in exercise physiology, and she suggested, "Why don't you go to med school?" It was the first time I'd thought about it. I'd never known anyone who went to med school. No one in my family had. Who did that? But as I thought about it, I could see it was the best choice.

Colette and I started doing some research. Some doctor friends thought I'd never get in, but fortunately, I had achieved a 4.0 GPA in college, and, despite my PE major, I had started out taking engineering courses, so I had plenty of science and math credits. I still had to take two semesters of organic chemistry and a biology course, and I knew I had to get an A in all of them. But after looking into it further, I thought I might be able to do it—and I did.

When I started medical school, I knew exactly what I wanted to do. At the time, no one was specializing in preventive cardiology, so I pioneered a course program for myself. I took every class I could that I thought would help me answer the question: What can we do to prevent heart disease? In the meantime, Colette and I got married and had our first son, Tyler. My wife and I graduated on the same day in 1983, she with a law degree, and I with a medical degree. After that, I had to choose a residency location. We set our sights on Portland, Oregon, where Phil

Knight, whose own son later died of an undetected congenital heart defect, had established his running empire, Nike.

Those early years on the banks of the Willamette River, under the shadow of Mount Hood, became the setting for many pivotal learning experiences that influenced my career as a preventive cardiologist. In the decades that followed, I travelled all over the United States and the world to work with renowned physicians and medical professionals, all united in our desire to end heart disease. For over forty years, I've worked as a leader in preventive medicine, including serving as a medical director at the National Center for Preventive Stress Medicine in Phoenix, Arizona; the medical director of the Preventive and Stress Medicine Center in Denver, Colorado; the director of stress medicine and hypertension at the world-renowned Cooper Clinic; and finally, the founder of the Boone Heart Institute in Denver, CO. Since 2008, our mission has been to completely eradicate heart disease and stroke. In the past fifteen years, we have grown, both in size and reputation, into a national and international destination for those seeking to detect the early signs of atherosclerosis, stroke, dementia, and other related diseases. The Boone Heart Protocol, which you'll learn about in this book, utilizes the most cutting-edge practices and technologies for heart imaging and genetic screening. We also specialize in the Mental Office Stress Testing (MOST) protocol, which I pioneered when we realized the tremendous impact that mental and emotional stress has on the heart (a topic we'll discuss more in Chapter 6). Thanks to these efforts, we've saved thousands of patients from life-threatening coronary events, and we now see many more thousands each year, always with the expectation that they will live free of atherosclerosis to ninety and beyond.

Thanks to my love of sports, I've had the opportunity to work with Major League Baseball, the National Football League, and, yes, the NBA. Throughout the book, I'll share many stories and the invaluable lessons I've learned from working with this high-risk group of elite athletes. In many ways, their special circumstances, injuries, and lifestyles provide the best testing grounds for advancements in preventive healthcare that can benefit everyone. My continued involvement in sports has also led to a growing concern with the alarming number of sudden, unexpected deaths among high school athletes. In 2012, the Boone Heart Institute founded EveryHeart, a nonprofit organization dedicated to providing America's high school athletes with free preventative cardiology testing to provide early detection of heart defects, arrhythmias, and other often unknown heart conditions. To date, we've tested over seventeen hundred students and found at least fifteen significant, life-threatening issues that required surgery. Thanks to early testing, these students' lives have been considerably lengthened, if not saved altogether.

One of the most exciting areas of development at the Boone Heart Institute has been the introduction of WAVi brain testing to identify risk factors for dementia, another disease of aging that we are learning is very closely related to heart disease. We're learning every day that the lifesaving techniques we use to save people's hearts and arteries just might save their brains, too.

What's Ahead

This book is the culmination of my decades of research and experience in the fight to end heart disease. I'm convinced that we have everything we need: the right testing, the right treatments, and the right tools to knock this disease off its platform as America's, and the world's, number-one killer. This guide offers curious heart owners the knowledge and resources they need to confidently advocate for their heart health, so that we all might live to be a healthy one hundred and beyond.

In Part I, we'll discuss the history and biology of heart disease, from what we know about it from an ancient Egyptian princess, to the four types of heart disease running rampant throughout our society today.

We'll examine how genetics and trauma predispose us to the disease, and how triggers like stress, smoking, chronic inflammation, and processed foods provide the catalyst to launch the disease onto its deadly path. We'll also explore the unfortunate reality of our current medical system's reactive approach to heart disease. The medical, mechanical, and technological interventions we've developed, such as stents, angioplasty, valve replacements, bypasses, and the like, to save victims of heart disease are no less than modern miracles; but they're only helpful in the final stages. Today's cardiologists are like quarterbacks, called in to make a fourth-quarter Hail Mary pass when the game is all but lost, and they're extraordinarily good at it. Columbia University Heart Surgery reports that coronary artery bypass grafting (CABG), one of the most common heart-saving interventions, has an overall success rate of nearly 98 percent. That's certainly far greater than Tom Brady's average touchdown rate of 5.4 percent. And yet, wouldn't it be better if we didn't need them at all?

Cardiovascular disease is the costliest disease that our healthcare system treats, producing an economic burden on our country that the Centers for Disease Control and Prevention (CDC) estimates at over $219 billion annually. Further, these fixes are only possible when we know about the problem, and most of the time, we don't. Often the first sign of heart disease is a heart attack, stroke, or death. These late-game efforts are valiant and successful when we can try them, but often symptoms arrive too late for treatments to be attempted at all. I believe that if we shift our approach to the disease, and, in doing so, shift the incentives of the healthcare system in how we treat it, we can completely eradicate heart disease and stroke as a meaningful cause of death within the next generation.

But how do we do that?

Part II offers very realistic hope for preventing heart disease—and not only that, but strokes, dementia, and other common conditions as well. We'll talk about how diseases are cured by looking at the incredible progress we've made as a society with once-deadly diseases like tuberculosis and AIDS as case studies, and how heart disease can easily follow the same path. In Chapter 9, you'll learn the Boone Heart Protocol—the simple, four-question method of inquiry used in our office to help us arrive at a diagnosis and treatment plan that is designed uniquely for each individual patient. I'll introduce the exact lifesaving tests and protocols we implement at the Boone Heart Institute, and how these same treatments help to reduce the risk of brain disease as well.

Throughout each chapter, in addition to hearing stories from my own experience, you'll meet a series of characters whose stories will help to illustrate how heart disease works, and how we treat it. These characters are fictional composites designed to protect real patients' identities throughout the book—but keep in mind, the challenges they faced and the solutions we designed are very real.

As I write this, I am seventy-two years old. I plan on playing basketball with my grandkids, and, God willing, my great-grandkids, for many years to come. The eradication of heart disease is not just a dream; it's inevitable. We *will* eradicate heart disease. The only question is when, and whether you and I will live to see this brighter world.

PART I

The Silent Killer

CHAPTER 1

A History of Heart Disease

Mark, fifty-four, had no idea when he woke up this morning that mere hours after his standard breakfast of coffee and a banana, driving his youngest daughter to high school, and commuting thirty minutes to the office, he would be fighting for his life. When our time comes, none of us do.

On this particular morning, at 9:37 a.m., Mark, an executive at a biotech company, is sitting at his desk in a corner office, contemplating a disappointing quarterly earnings report and dreading this afternoon's presentation to the board. He recalls the argument he had with his wife last night and the struggle to get his daughter, who slept in, out the door on time. He arrived later than usual and feels rushed to try to explain his department's poor sales. With all of this weighing on him, Mark is stressed, and though his concerns are less visceral than the threat of an angry warthog or an enemy archer would have been to his ancestors, they are no less real to him, or to his physiology.

Mark has always taken care of himself. He runs or lifts weights five days a week. He takes his family on an annual beach vacation to relax. He reads before bed because he once read that one should not look at screens for the last hour of the day. He is close with both his siblings, who live nearby, and has developed deep connections with his community through nonprofit work. He sleeps seven hours each night. Most days, he feels good.

But despite his workouts and his standing desk, Mark lives a more sedentary existence than our hunter or farmer predecessors could ever have imagined. Each day, for more than a decade, Mark has driven eleven miles to the office and eleven miles home. In his downtime, he goes to the movies with his wife, or watches football from his couch. Mark tries to eat a diverse diet of primarily lean meats, unsaturated fats, and vegetables, but age has slowed his metabolism and he can't always resist finishing his children's leftover pizza or French fries. He enjoys a glass of wine most nights, and more than one on the weekends. Mark has never smoked.

Most crucially for our purposes, Mark has particular genetic tendencies that make him especially vulnerable to certain diseases. While he has his basic cholesterol and blood pressure checked every few years at his sporadic checkups, his doctor has never checked for the genetic risk factors that predispose him to a variety of deadly diseases. Most doctors don't.

Likewise, Mark's heart and his arteries, which have been working fine all his life, as far as he knows, have never been examined with any sort of imaging. If it were as routine to check for plaque in our hearts and blood vessels as it is to check for plaque on our teeth, the sonographer would have discovered a chunk of soft plaque growing in Mark's carotid artery. If Mark's physicians had been prompted to image his heart with CT angiography, a painless and safe test, they would have discovered a "widowmaker" blockage in his coronary artery, covering 40 percent of the artery's diameter.

These vulnerabilities have always existed in humans, but only in the last decade or two have we been able to identify them. These tests are often considered unnecessary in a patient like Mark, but they can be crucial in identifying the supposedly healthy patient who is on the brink of disaster. Since his twenties, deadly pools of cholesterol have been gathering in the linings of his arteries. This process unfortunately happens in all of us as we age. Cholesterol particles are around the same size and shape as streptococcal bacteria, and so the body views them as invaders and attacks them once they lodge in the artery's walls. This causes pockets of plaque to develop, gathering under the artery's inner lining like pools of lava, ready to erupt—a process we'll learn more about in Chapter 2.

On this day, one particular deposit in Mark's coronary artery is particularly vulnerable. Mark's stressful morning has elevated his blood pressure, forcing blood through his arteries faster and harder, including in the fragile coronary artery. All at once, the plaque ruptures. The volcano bursts. Mark feels the telltale pain in his chest, and experiences shooting pain and weakness in his left arm.

Mark's body rushes to save him. Platelets flood through his bloodstream to the site of the rupture, gathering there and forming a clot to "heal" the wound inside his artery, just as your platelets would gather to form a scab over a skinned knee. Only, instead of stopping him from bleeding out, this clot only worsens the clog in the central artery of his chest. Instead of saving him, it's killing him.

The Lesson of the Princess

Despite the portrayals of heart attacks in the media, at least 45 percent of heart attacks occur in patients like Mark, with no prior symptoms. That means no chest pain, no shortness of breath, and no irregular heart rate. Furthermore, 70 percent of all heart attacks occur in patients who would be considered "low risk" by current heart health standards. We tend to think of heart disease as a consequence of extreme excess and poor lifestyle habits; but as you'll see, this is an oversimplification. Yes, lifestyle is a contributing factor, but heart disease is a much more complex and nuanced adversary than we've given it credit for. In fact, it may even be dangerous to think of heart disease as a disease of modernity and/or of plenty. It serves no one to think of the disease most likely to kill all of us as something that preys only on the lazy or the obese. By definition, the leading cause of death preys on everyone. We see the truth of this in the way it has crept up on us, striking with deadly intent, throughout history.

The first documented cases of atherosclerosis, the most common type of cardiovascular disease, in which the arteries calcify over time, were discovered in a number of ancient Egyptian mummies. Using CT scans, in a 2011 study, medical researchers were able to identify the extent of the damage to gain new insights about this not-so-modern disease. Of the forty-four mummies that still had an intact heart and arteries, twenty had recognizable signs of calcification throughout the blood vessels and aorta. One mummy in particular showed extreme atherosclerosis, with widespread calcification throughout the entire body. The body belonged to an Egyptian princess, who lived between 1580 and 1550 B.C. and is thought to have died sometime in her forties. This unexpected discovery signaled that heart disease was more common in ancient times than previously thought.

We can draw some reasonable conclusions about why this princess and other Egyptians may have been vulnerable to atherosclerosis. In ancient Egypt, mummification was time-consuming and expensive, and therefore reserved for the wealthy. Perhaps the princess lived a more sedentary life than the average person of her era. The typical diet in ancient Egypt was rich in vegetables and whole grains, what we would consider today fairly "healthy," but it is likely that the princess had regular access to calorically richer foods like meats, butter, and cheese as well. However, with the knowledge we've gained in the past decade about genetics, we can also venture that the princess may have been genetically predisposed to the disease, and that these genetic factors, combined with her more lavish lifestyle, hastened her passing at quite a young age (we'll talk more about genetics in Chapter 3).

In fact, I would be willing to bet that the princess's chefs and servants, and even the workers building pyramids in the heat of the desert, all had precursors of heart disease, as well. One remarkable finding from the study was that all the Egyptian mummies who showed signs of atherosclerosis died in their forties, while the ones *without* signs of atherosclerosis died much earlier, in their mid-thirties. My guess is that, given enough time, atherosclerotic calcification would have accelerated in the arteries of even those younger mummies. Something else just killed them first.

The Rise of Our Number-One Killer

For most of our species' history, humankind has been fighting to survive. Between food scarcity, predators, harsh weather, and natural disasters, our first ancestors were lucky to make it to age thirty. In the ancient hunter-gatherer societies that existed for millennia, our *homo sapiens* ancestors had more immediate worries than things like heart disease or stroke, and the fossil record reflects that. In fact, the threat of attack from other humans and animals, like the saber-toothed cat, ancient wolves, and crocodiles forced our bodies to evolve the lifesaving mechanism of sending platelets to scab and heal exterior wounds—the same mechanism that now causes life-threatening blockages in our hearts. Dying of one of what we now call the "diseases of aging"—heart disease, stroke, cancer, or diabetes—would have been quite an anomaly. Indeed, such a fate would have represented the end of a tremendously successful life for these earliest humans.

Around twelve thousand years ago, during the first agricultural revolution, people gathered in communities and domesticated certain plants and animals, paving the way for much more complex human societies. The basic building blocks of survival—food, water, and shelter—could now be provided for by the collective. These societies saw the first surpluses of food, which led to a population boom and new occupations for people whose lives were not solely dedicated to producing those basics of survival. These new occupations included artisans who specialized in skills like carpentry, masonry, pottery, basket-weaving, and weaponry. It introduced the need for organization and hierarchy, in the form of lords, landowners, and princesses. Agriculture also allowed for the profession of medicine to come into existence. Hunter gatherers likely knew how to stanch a wound or use local plant life to treat a burn or an insect bite, but evidence suggests that they did not have specialized physicians. Back then, everyone was a jack-of-all-trades, doing whatever it took to survive. Settlements and food surpluses allowed for the emergence of specialized doctors, who got to work addressing their patients' physical needs—of which there were plenty.

While food generally became more readily available, overreliance on specific crops, such as grains, made these fledgling communities vulnerable to new environmental and biological threats. They routinely suffered from crop blights, famine, and drought, which in turn led to starvation and social inequalities. In addition, the nutritional diversity of their diet decreased, leading to an overall deterioration of health. Substantial evidence suggests that humans decreased in height, experienced tooth decay and cavities, and encountered new infectious diseases. The increase in population density was a boon to viruses and infections, which ran rampant. Think of the Black Plague, which in the fourteenth century killed more than one-third of Europe, or twenty-five million people, in just four years—an unimaginable number to our modern minds.

For centuries, we died of accidents, violence, illness, and infection. If you survived childhood (by far the most vulnerable years for our predecessors), you might at best expect to live another twenty years, into your late thirties or early forties. Doctors learned what they could about the way we died, and made great and invaluable discoveries, but they arrived at few actual *cures*. It was not until the late nineteenth and twentieth centuries that medicine truly began to win some battles.

In the early 1900s, cardiovascular disease competed with pneumonia, tuberculosis, and influenza for the ranking of number-one killer. The Spanish flu in 1918 caused the deaths of at least fifty million people, and marked the last time an infectious disease outranked heart disease as the leading cause of death. The greatest leap forward came with the discovery of penicillin in 1928 by Alexander Fleming, an event which likely saved the most human lives in history. With the development of antibiotics, infection became manageable and gradually fell down the list of common causes of death. Later, development of vaccines and inoculations gave us tools to fight viruses on a microscopic level. In first-world countries, human lifespans rapidly doubled over the course of a century.

But as medical advancements have largely helped us fight and keep at bay communicable diseases, heart disease has moved in to take their place. As we saw with the Egyptian princess, this slow-moving disease has always been within us, but never got a chance to move to the front of the line until the scientific revolution of the last century wiped out the competition.

Of course, our increasingly sedentary lifestyles and consumption of chemically engineered "hyperpalatable" foods have only accelerated this process. In some ways, we all live like Egyptian princesses now. Humanity has ascended to the top of the food chain so quickly that our bodies have not had time to evolve again. While we have advanced from hunters to farmers to office workers in a relatively short time, we retain the biology of our hunter-gatherer predecessors. So, when we have a heart attack, based on millions of years of evolution, our bodies react like we've been attacked by a mountain lion or stabbed with a spear. In other words, as our societies have gotten *better*, the disease has gotten worse.

The Fight to Save Our Hearts

By the 1940s, death by heart disease was viewed as almost an inevitability. I recently read a book about the administration of Franklin Delano Roosevelt, which ran from the Great Depression through World War II. While enjoying a deep dive into this tumultuous but ultimately triumphant period in our nation's history, I was struck by the fact that almost every month another member of Congress or a high-ranking government official died of a heart attack or stroke. Indeed, these events became little footnotes to the sweeping events of history. Clearly, something needed to change; the problem wasn't going away.

Even President Roosevelt himself was a victim. As Drs. Syed S. Mahmood, Daniel Levy, et al, report, there were signs during his administration that the president was in danger. They write, "In 1932, candidate Roosevelt's campaign office released medical records showing his blood pressure to be 140/100 mm Hg, which did not prompt any medical intervention. [...] Between 1935 and 1941, the president experienced a gradual rise in blood pressure from 136/78 to 188/105 mmHg." For contrast, a healthy blood pressure is typically less than 120/70 mmHg. Despite these warnings, FDR's doctors declared that he was healthy. It wasn't until March 1944 that his daughter Anna requested a second opinion, and Roosevelt was seen by Dr. Howard Bruenn, who diagnosed him with "hypertension [high blood pressure], hypertensive heart disease, and cardiac failure." The president was treated with digitalis, a common drug thought to increase circulation, and was told to stay off the salt. Less than three months later, Allied troops landed on the beaches in Normandy, ushering in the end of World War II. The following year, in April 1945, Roosevelt died of a stroke with a blood pressure of 300/190 mmHg.

In 1948, after Roosevelt's passing, President Truman signed the National Heart Act, which established the National Heart Institute and provided funds for an extensive study on heart disease. In fact, plans for the study were already in place. Dr. Joseph Mountin, an expert in the epidemiology of communicable diseases, had recently founded the Communicable Disease Center, as it was then known. Today, we know it as the Centers for Disease Control and Prevention (CDC). Despite his background in communicable diseases, Mountin had become interested in studying chronic, noninfectious diseases and was concerned about the rise in heart disease. He proposed the idea of a longitudinal study of heart disease, the first of its kind, and chose Gilcin Meadors to lead the twenty-year study.

The researchers chose Framingham, Massachusetts, as the site of the study for its proximity to Harvard Medical School, and because the town had participated in a similar study of tuberculosis in the late 1910s. Dr. Paul Dudley White, a star cardiologist at Massachusetts General Hospital, was also on the Framingham team. They examined and surveyed nearly six thousand Framingham residents, and followed up five to ten years later. The first results of the study revealed the presence of certain "risk factors" that made patients more susceptible to heart disease. This is common knowledge to us now, but it was revolutionary at the time and directly contradicted the assumption that heart disease was unavoidable. Instead, Framingham proved that several lifestyle factors influenced the disease, meaning that people's actions had some influence and control over the outcomes of their health. For the first time, individuals realized that they may have some power in the battle against this deadly foe.

Just a few years after the first Framingham results, in 1955, four years after my birth, President Dwight Eisenhower suffered a massive heart attack on a golf course in Denver, Colorado, not far from where my family would settle almost half a century later. Thankfully, President Eisenhower did not die from this unexpected event. Indeed, the innovative steps that his presidential medical team took following his heart attack would, in some ways, change treatment forever.

In the 1950s and 60s, there was no agreed-upon treatment for a heart attack. If, like President Eisenhower, the victim survived the initial event, they were given painkillers and confined to a hospital bed indefinitely. Bed rest was the only prescription available.

You can see how the medical community at the time came to this conclusion. Time and time again, a man or woman, who appeared completely healthy and pain free, was suddenly stricken with a heart attack or a stroke, disabling or killing them without warning. After their cardiac event, these victims of seemingly random events, must have seemed fragile and unstable. Without the technology to see what was going on in their heart, their arteries, or their bloodstream, each patient must have seemed like a ticking time bomb, ready to go off at any slight disturbance or shift of the wind. Best then to leave them undisturbed and immobile.

Enter Paul Dudley White, MD. Following his heart attack, Dwight Eisenhower's heart health was placed in the hands of Dr. White, who had recently won an award for his participation in the Framingham study, and had helped to found the National Heart Institute. Dr. White's approach differed from the conventional wisdom at the time in important ways, and history now remembers him as the founder of preventive cardiology. Dr. White strongly believed that lifestyle had a powerful effect on coronary heart disease. Over the years, he had become a staunch advocate for exercise, diet, and weight control as cornerstones in the prevention of heart disease. So, when Dr. White walked into Eisenhower's hospital room, one of his first orders was for the president to begin walking—and even cycling!—as soon as was physically possible.

More conservative interventional cardiologists must have been aghast. Dr. White's recommendations were the opposite of the protective, infantilizing strategies that had been used. But they made sense. Consider a broken bone. Following a fracture, the impacted limb—let's call it a broken leg—must be immobilized until the bone can knit itself back together. But as those of us who have broken our legs know, physicians recommend weight-bearing and immediate activity as soon as possible, pain notwithstanding. This spurs the bone to re-form more quickly, as well as rehabilitating the muscles around the injured area. Strict rest leads to atrophy, which can then lead to more problems throughout the musculoskeletal system. The same is true for our hearts.

Dr. White's aggressive approach would take decades to become the standard of care, but it was the start of a shift away from reaction and towards prevention. Throughout Eisenhower's recovery, Dr. White regularly informed the media of the president's progress. When the press began to wonder if the heart attack had eliminated Eisenhower as a candidate for a second term as president, Dr. White publicly declared that running for a second term was not just possible, but essential to the president's recovery.

The word *essential* sticks out to me as a bedrock of Dr. White's revolutionary philosophy. Living our lives to their fullest, even after a traumatic event, is essential to recovery. White practiced what he preached, once walking from Washington National Airport to the White House for a consultation with President Eisenhower. Until his death, Dr. White traveled the world, preaching prevention through diet, exercise, and weight control. President Eisenhower—coincidentally another Kansan, raised on the endless prairies in Abilene, an even smaller town than Garden City—recovered spectacularly and served as president through 1961. Incidentally, future president Lyndon Johnson had suffered his own heart attack five months before Eisenhower's while serving as the Senate majority floor leader. News of Eisenhower's recovery and Dr. White's counterintuitive methods reportedly inspired Johnson to return to the Senate later that same year. The winds were gradually changing, but getting presidential-level healthcare to the masses can take decades.

Pioneering Preventive Cardiology

By 1985, when I was a second-year resident at Good Samaritan Hospital in Portland, Oregon, preventive heart care had yet to take root even in medical circles. Heart attack victims came in all the time and often died in the hospital under our care. Plain Old Balloon Angioplasty (POBA) was the most common treatment at the time, involving the placement of a balloon at the end of a long catheter into the artery to widen its diameter around the clogged area. Elastic recoil—that is, the artery retracting and becoming clogged again once the balloon was deflated—was a common problem. Within the next year, the first coronary stent was developed to prevent this recoil and was approved for use by the FDA in 1987. However, stents themselves were faulty and led to other complications. (Don't worry; these procedures have significantly improved since they were developed in the 1980s!)

As you already know, often by the time cardiovascular disease has been identified (likely because of a heart attack or stroke), it's too late, though this didn't seem to disturb my colleagues very much. Compared to the other diseases we witnessed at the time, such as the first AIDS cases in the Pacific Northwest, cardiovascular disease seemed a much gentler way to go. The doctors and nurses did their best to make our cardiac patients comfortable, but effectively shrugged their shoulders and shook their heads. I heard them saying, "He should have smoked less," and, "She should have eaten better." The subtext: what more can we do?

I was convinced there was plenty more we could be doing; I had already seen evidence of what was possible during my time at the YMCA. My residency program was spearheaded by Dr. Stephen R. Jones, an infectious disease specialist, now tasked with supervising the care of patients suffering from the greatest infection the world had ever seen: HIV. But despite the hospital's preoccupation with fighting HIV/AIDS, Dr. Jones was supportive of me pursuing my own interests. That year, I was made the director of cardiac rehabilitation at the hospital, and I set about attempting to bring more of the lessons I had learned earlier in my career to helping patients recover post-cardiac events. Stents and bypasses, which might rescue ten inches' worth of blood vessels, were occasionally useful for stalling the progress of the disease for a short time, but we have over sixty thousand miles of blood vessels in our bodies. I knew that much more needed to be done in order for patients to avoid a second, and probably more deadly, event. The problem was that the disease is invisible on the outside, and we didn't have the drugs or the imaging to be able to effectively see inside.

Coincidentally, researchers at Stanford, led by Dr. Robert DeBusk, were exploring the efficacy of physical conditioning and exercise after myocardial infarction (heart attack) at that time. Their findings supported what Paul Dudley White had suggested and what I myself had observed at the YMCA: that moderate exercise after a cardiac event improves health outcomes for these patients and decreases the likelihood of another infarction. DeBusk and his team had developed a revolutionary monitoring system where participants in studies who were recovering from a heart attack could go about their normal lives, including their normal exercise routine, and record electrocardiogram (ECG) data virtually. In the pre-Internet era, this was the height of technology. My supervisors in Portland were interested in adopting the same technology and practices, so that we could monitor our own rehabilitating patients. I was sent to Stanford to learn from Dr. DeBusk for a few days. The technology was a game-changer for us, and anyone who'd had a heart attack at Good Samaritan was enrolled in the program. Our patients could be hiking 250 miles away in Crater Lake National Park, and we could still easily monitor their exercise and tell them if their ECG showed that they were safe.

We now knew with confidence that we could help patients recover and heal from a cardiac event. Naturally, I started to wonder: If we could help them heal through imaging, virtual monitoring, and exercise, could we help them avoid an initial event in the first place?

This question led me to the next phase of my career. When I completed my residency, I thought about becoming a board-certified interventional cardiologist, which would have taught me how to perform the Hail Mary procedures I had been working to prevent. It was certainly a lucrative option. It also appealed to me because the technologies around stents and bypasses were developing rapidly, and those procedures require incredible skill; it was an exciting time to pursue cardiology. Ultimately, though, I felt there were already enough interventional cardiologists choosing that route. I decided instead to join the ranks of pioneers like Paul Dudley White in the budding field of *preventive* cardiology, the far less sexy option. It was both challenging and ambiguous, primarily because the disease I was fighting was asymptomatic. People are starting to recognize today what I learned in my residency program forty years ago: that if you wait for symptoms of heart disease, you've waited too long. By the time you feel pain in your chest, you've already had the disease for thirty or forty years, and it may be too late.

Perhaps it's because we've had to fight so hard for the privilege of dying, not from animal attacks or infectious diseases, but rather from a "disease of aging" that we tend not to take heart disease as seriously or fight it as urgently as other more obvious and ostentatious causes of death. And yet, as you'll see in the next chapter, it is precisely this ability of heart disease to hide in the innermost parts of us and escape detection for decades that makes it so painful, deadly, and, yes, heartbreaking.

CHAPTER 2

Building Plaque: Atherosclerosis

It's a warm July day in Hardwick, Vermont, 1984. A lone jogger is running on Route 15, an escape from his motel to get some fresh air and exercise while traveling. This is what he loves; he believes in the power of running to cure any ailment. He often runs ten or more miles per day. Just the day before, he walloped his partner in a tennis match. He's fifty-two, and at the top of his game—or so he thinks.

A few hours later, a passerby finds his body lying on the side of the road.

The runner was Jim Fixx, the author of *The Complete Book of Running*, who popularized the idea of jogging in the 1970s and 80s. Fixx was known for pushing himself. One of his favorite shirts said, "You haven't really run a good marathon unless you drop dead at the finish line."

I imagine when he set out on his run that day, he had no idea how prescient these words would be.

Ironically, Fixx first took up running in his thirties after a tennis injury because he was concerned about his family history of heart disease. His own father, Calvin, had had his first heart attack at just thirty-five years old, and a second one that killed him at forty-three. Fixx had made it past these important age milestones, and believed that his running had saved him from the same fate as his father. He stopped smoking, lost weight, and was persuaded that running had the power to change lives. However, even the indefatigable Jim Fixx could not stop his genetics from catching up with him at age fifty-two. The cause of death was a massive heart attack while jogging. Autopsies showed multiple blockages in his arteries—blockages that undoubtedly would have warranted bypass surgery to restore blood flow to his heart, had they been caught in time. Better yet, these blockages could have been avoided altogether if Fixx had known what we know now about atherosclerosis. He would have been ninety-one in 2023, and perhaps still jogging.

There have been many interpretations of Jim Fixx's death. For some, it's a convenient cautionary tale about the dangers of exercising too much. For others, his survival to age fifty-two—nearly ten years beyond his father!—is evidence that his running really did prolong his life. Both explanations have some truth to them, as we'll discuss in later chapters. But I see his story another way altogether.

No matter how healthy you appear to be, no matter how strong you truly are, heart disease can still end it all. And atherosclerosis, the type of heart disease that Jim Fixx had, is the most insidious—the kind that most people never see coming. It is our kryptonite—worse than that, actually, because most of us don't know it is our greatest vulnerability.

In the next four chapters, I'll introduce you to the four common forms of heart disease. You can imagine the heart like a house. You need to have solid plumbing, where the pipes are strong and clean (atherosclerosis—this chapter) with no sudden clogs present (thrombosis—Chapter 3). You also need a strong structure with doors that open and close effectively (the muscle and valves—Chapter 4), and proper wiring to keep all the lights on (electricity—Chapter 5). After decades of working with seemingly healthy patients, I've learned that no matter how good the exterior of the house looks, if something is wrong in one of these areas, those issues need to be addressed before the whole house collapses. In this chapter, we'll start by discussing the most common and deadly risk to the heart: atherosclerosis. As you'll see, atherosclerosis grows subtly for decades in the lining of our blood vessels, often as a result of genetic factors accelerated by poor housekeeping. Like stockpiling dynamite in the basement, atherosclerosis waits for the perfect storm of circumstances that will light the fuse and send the whole structure tumbling down.

The Delicate Balance of Cholesterol

Remember Mark from Chapter 1? Despite his relatively healthy lifestyle and seemingly low risk for heart disease, his coronary artery was riddled with plaque, compounded by decades' worth of cholesterol buildup and inflammation in the fragile lining of his arteries—until, one stressful morning, the plaque ruptured. Neither Mark nor his doctors saw it coming. In the last forty years, researchers have been hard at work to understand this complex phenomenon and how it occurs, often apparently out of nowhere.

We have roughly sixty thousand miles of blood vessels throughout our bodies, through which our blood transports all the oxygen, nutrients, proteins, and hormones we need to our muscles, tissues, and organs. In fact, the blood vessels are the largest organ in the body, long enough to wrap around the earth nearly two-and-a-half times. If you sliced all your blood vessels in half lengthwise and laid them out flat, they would cover the expanse of a tennis court. Our arteries and veins are the superhighways of the body, enabling quick communication and mobilization of our immune system, eliminating waste, and delivering whatever materials we need for life to sustain itself via the blood. We often think of blood almost as a passive river of nutrients and oxygen, but "blood is a network," author Siddartha Mukherjee writes. It is a highly complex communication pathway where the cells within our bodies convey the information and materials necessary to sustain life.

One of these materials is, of course, cholesterol.

Cholesterol has been widely misunderstood and maligned, although we're beginning to see a realignment in public understanding of this important element in the body. It's too simplistic to say, as many of us were taught in decades past, that there is "good" or "bad" cholesterol, or that cholesterol is solely responsible for atherosclerosis. In fact, cholesterol is an essential type of lipid, or fat, refined and purified for its many uses in the manufacturing centers in the liver. Our bodies rely on cholesterol for a number of important functions, such as digestion and maintaining the integrity of our cell structure. Cholesterol is also involved in processing and regulating stress hormones like cortisol, and sex hormones like estrogen, progesterone, and testosterone.

The role of cholesterol in the progression of atherosclerosis was first confirmed by the Russian doctor, Nikolai Anichkov, in 1913. As Meyer Friedman and Gerald W. Friedland explain in their book *Medicine's 10 Greatest Discoveries,* in 1908, while Anichkov was still a medical student, a clinician at his school named A.I. Ignatowski had the idea of trying to generate atherosclerosis in rabbits in order to observe its effects. His experiment was successful; after a few weeks of feeding the rabbits a diet of eggs and milk, he found atherosclerotic plaque in their aortas. But Ignatowski incorrectly theorized that it was the *protein* in the milk and eggs that had caused the plaque. He moved on to other scientific endeavors, but his discovery ignited a flame of curiosity in Anichkov.

Now a recent graduate of medical school, the young doctor Anichkov, and a couple of other medical students, decided to replicate Ignatowski's experiments on rabbits and study the atherosclerotic plaques that accumulated in their aortas. The students observed that the plaque had fatty droplets in it, and that those same droplets could be found in the rabbits' livers as well. Knowing that cholesterol produces the same fatty cells, they conducted the experiment again, but instead of feeding the rabbits eggs and milk, they fed them only egg yolks, hypothesizing this time that it was the cholesterol in the yolks that caused the atherosclerotic plaque his colleagues had observed. They were correct; again, after a few weeks, they found a buildup of plaque throughout the rabbits' arteries. Many scientists took this as a clear sign that dietary cholesterol was the sole cause of our most deadly disease. For decades the dominant theory of atherosclerosis prevention was to stop eating cholesterol-rich foods like egg yolks. It's not bad advice, but for most of us, as for Jim Fixx, not eating egg yolks won't make much of a difference.

Still, we have Dr. Anichkov and his associates to thank for setting us on the path to learning more about the complex role that cholesterol does play in the development of atherosclerosis.

Cholesterol is carried in the bloodstream by lipoproteins, which dutifully deliver the cholesterol (and other nutrients) to our bodies for use. If the blood vessels are the highways in our body, then imagine the lipoproteins as individual cars shuttling their passenger, cholesterol, around like Uber. Our Uber lipoproteins are supposed to drop their cholesterol passengers off, then park within the factories in the liver, awaiting more raw materials for manufacturing. The problem today is that most of us have too many Ubers clogging our interstate freeways!

Building on Anichkov's work, in 1950 Dr. John Gofman discovered that there are two different kinds of lipoproteins: low-density lipoproteins (LDL), which deliver the cholesterol to cells, and high-density lipoproteins (HDL), which absorb the excess, unused cholesterol from the system and transport it back to the liver, where it is received by LDL receptors and removed again from the bloodstream. In this way, our lipoproteins and the liver are constantly monitoring the amount of cholesterol in our system to keep it in check and make sure we have just the right amount circulating in our bodies. Ideally, the cars bringing cholesterol (LDLs) are evenly paced with the cars transporting cholesterol away (HDLs).

The problem comes when there is an imbalance between the two types of lipoproteins, or when the LDL receptors in the liver aren't functioning—in other words, when there are too many cars delivering cholesterol around, and not enough to take it away. When this happens, the number of LDLs delivering cholesterol starts to outpace the HDLs transporting it to the liver for removal.

We've all learned to recognize certain vehicles by markers they have that identify their function in society: police cars, taxis, mail carriers, and fire trucks are all clearly marked to tell us exactly what their purpose is. Similarly, our lipoproteins have markers on them called *apolipoproteins*, proteins that give them their structure and are critical in the functioning and metabolism of our cholesterol. These apolipoproteins announce to the surrounding cells what type of lipoprotein this is, and what it does. HDLs bear the marker of apolipoprotein-A1 (apoA1), which increases their capacity to transport the excess LDLs. They're like the vacuum cleaners or garbage trucks of the blood, taking cartloads of LDLs back to the liver for "processing." LDLs, on the other hand, have markers called apolipoprotein-B (apoB). ApoB can actually be found on a number of different types of cells, including LDLs, and none of them are good. Scientists now suspect that it's not the amount of cholesterol-carrying LDLs alone, but actually the total amount of ApoB-bearing cells in your body (on both LDLs *and* other cells) that determines how likely you are to have atherosclerosis. This is a more accurate reflection of your overall cholesterol load on your system, and is becoming the standard way of measuring the presence of plaque. Doctors have found that patients with atherosclerosis tend to have an elevation of both apoB and LDL levels, overwhelming the apoA1-carrying HDLs. As cardiac electrophysiologist and cardiologist, Dr. Mahnaz Behbodikhah, et al., wrote: "The ratio of apoB/apoA1 reflects the balance between the two opposing forces and has been proven to be an accurate indicator of CVD [cardiovascular disease] risk. The greater the ratio, the more cholesterol is circulating in the plasma and being deposited into the arterial wall."

Glagov and Cream Donuts

After Anichkov's discovery, most doctors believed that excess LDL simply clogs up our blood vessels, much like excess hair or food clogs up the drains and pipes in houses. It makes intuitive sense. But that notion was challenged in 1987 when Dr. Seymour Glagov at the University of Chicago made a discovery that changed our understanding of atherosclerosis forever.

In what was soon to be a groundbreaking study, Glagov autopsied the hearts of 136 patients who had died of heart conditions in order to assess the extent of the damage from plaque within their arteries. What Glagov found is that the plaque in our blood vessels doesn't accumulate the same way it does in the pipes of a house. This plaque doesn't grow *inside* the blood vessel, blocking the blood flow (at least not at first); instead, it grows in the wall of the blood vessel itself, underneath the *inner lining* of the vessel, called the endothelium. The first two levels of this blood vessel lining where atherosclerosis begins is called the intima media complex. This lining is like the pavement stretching along the entire sixty thousand miles of blood vessels that form the highways in your body.

The intima media plays a central role in blood flow regulation. Importantly, it is composed of the endothelium, a semipermeable layer, just one cell thick, which functions much like a sieve to allow essential nutrients to be absorbed into the wall of the blood vessel and keep the blood vessel flexible and pliant. I like to call it the "yellow brick road" as a nod to *The Wizard of Oz,* in honor of my Kansan roots. Measuring the thickness of the intima media is one of the most important tests we run at the Boone Heart Institute, because the health of the intima media gives us the most accurate picture of you at the time of the test. The intima media tends to be thinner in women than men, and thinner when you're not stressed. It also tends to be thinner when you have low inflammation, low cholesterol, and low blood pressure. Everything that can make your body worse usually gets measured in that yellow brick road. The thicker it is, the more action we need to take.

Unfortunately, the permeable endothelium also leaves the intima media vulnerable to excess LDL particles that might be floating around carrying cholesterol. LDLs are just small enough to slip through the endothelium and stack up in the wall of the vessel—in fact, they are about the same size as streptococcal bacteria, the same bacteria that causes strep throat and pneumonia, and they damage the endothelium in the same way, weakening the strength and thickness of the lining, leaving it vulnerable to even more LDLs coming through. Imagine if your blood vessels had strep throat; that's essentially what's happening here.

When the LDLs make it inside, the wall doesn't respond by pressing inward; it first responds by expanding the vessel *outward*. Because of this, the plaque doesn't affect blood flow for a long time, because the opening within the blood vessel where the blood flows—called the lumen—remains the same size. Instead, it just creates little road bumps on our superhighways. Annoying, yes, but ultimately not impeding traffic. Yet.

Imagine a round donut being injected with a cream filling. The cream goes into the outer ring, or the "wall," of the donut, and that wall expands outward to make room for it. Meanwhile, the hole in the center of the donut remains the same size. The lumen typically isn't compromised at all until there's a 40 percent cross-sectional filling of the artery wall with plaque, as shown in Figure 2.1. In other words, our arteries have the ability to expand up to 40 percent to accommodate plaque before any blockage occurs that's extensive enough to be noticeable on an angiogram. If a plumber looked at these pipes, he'd shrug his shoulders and pronounce, "Looks clear!" Similarly, nuclear medicine treadmill and stress echo treadmill tests won't pick up on plaque at all before the 40 percent point, because until then there's no significant blockage occurring. In fact, it usually takes at least a 70 percent luminal blockage to make these tests abnormal.

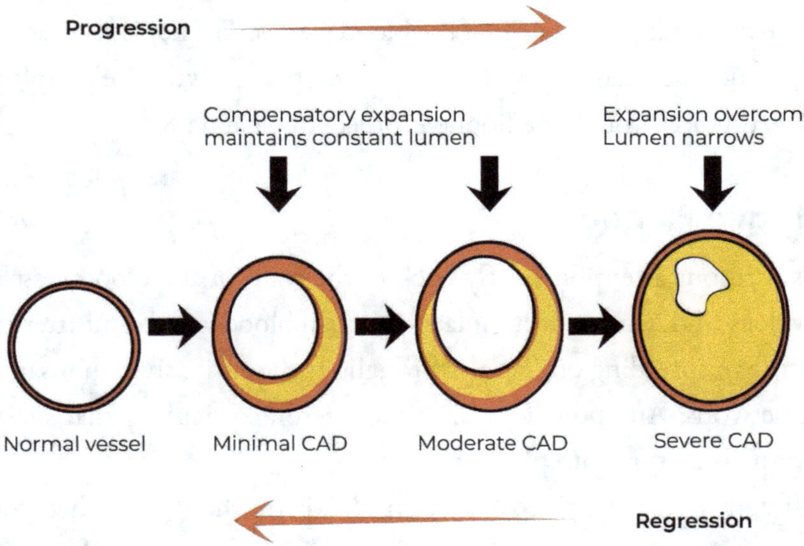

Figure 2.1

After the cross-sectional filling of the artery wall with plaque has passed the 40 percent (moderate) mark, the blood vessel is overwhelmed to the point that it finally starts collapsing inward—enough to be seen on an angiogram. This begins to narrow the passageway for blood flow. This is usually the ticking time bomb of an impending heart attack or stroke. However, most cardiologists won't be alerted to the danger until the lumenal tunnel is narrowed by 70 percent (severe).

This is our central problem: most people die *before* they show signs of serious blockages. Somewhere between 60 percent and 80 percent of heart attacks and strokes occur from blood vessels that have not yet reached that 40 percent mark, and aren't obstructed enough to cause chest pain, or even show up as abnormal on angiograms, giving doctors no chance to react with preventive measures. We see people like Jim Fixx, who are running marathons, climbing mountains, or passing treadmill tests one day, and then dropping dead of heart attacks the next. That's because most testing has focused on, and continues to focus on, the lumen and its current blood flow, when what we really need to look at is what's hiding in the walls of the blood vessels. The key is to discover any plaque, and then stop and slowly reverse that growth. We'll explore how to do this through the Boone Protocol in Chapter 8.

Hot Lava Plaque

Let's turn our attention briefly back to the walls of the blood vessels. Why does LDL tend to accumulate in certain blood vessels and arteries, rather than spreading out evenly throughout the entire sixty-thousand-mile network? And how does LDL cholesterol, particularly that sticky lipoprotein(a), turn into plaque?

If you've seen the videos of "trash islands" in the Pacific, then you may have an idea of how ocean currents tend to carry trash and plastics along until they collide with other currents, causing the trash to build up in dense "islands" of microplastics in the ocean—and, unfortunately, wreaking havoc on the marine life in that environment and disrupting fragile ecosystems. Blood operates in much the same way; our "trash," or excess LDL, tends to build up in the bends and branches of our arteries, where the blood flow is forced to change direction or slow down. There, the tiny particles of LDL slip inside the porous endothelium and begin a process of inflammation and oxidation—the two primary causes of all disease.

In the early eighties, when my family and I moved from Iowa to Portland, Oregon, for my residency at Good Samaritan Hospital, we packed up our old, beat-up Datsun B210 and headed west. We were coming from the harsh winters of Iowa to the chronic rainstorms of the Northwest, where moss practically grows on the concrete. The old Datsun was already beginning to rust, a speckled sheen of orange at first dotting the hood, the roof, the trunk, and even the undercarriage. Then, exposed to the rainy weather of the Pacific Northwest, an entire hole rusted through the floorboard. I remember being able to look down through the floor and watch the asphalt zooming by as I drove my infant son, Tyler, around town. It got so bad, in fact, that one day when I was driving to work, the whole bottom of the car dropped out. We finally sold it to a neighbor for $300, and he eventually left it somewhere on the side of the road to let it dissolve in the elements.

The same thing happens within the blood vessels as the strep-like, apoB-bearing LDL piles up and begins to oxidize, creating a lesion and weakening the integrity of the endothelium. In response, the body sends a specific type of white blood cell called a monocyte to try to handle the damage. The monocytes enter the intima media through the yellow brick road, thinking they're going to kill the cholesterol cells (in the same way our immune system would heal from strep throat or pneumonia). The monocytes engulf the LDLs, turning into larger immune cells called macrophages.

The macrophages initiate an inflammatory response to the oxidized LDL particles in an effort to kill the rogue cells. This is what inflammation is; the body is literally manufacturing heat to kill the wayward bacteria—just like when you have a virus and your body generates a fever to kill it. Unfortunately, however, the macrophages aren't effective here; they're no match for the LDLs. Instead, they stack up, coalescing into large, hot "foam cells" that create plaque. Over time—decades, for most of us—the buildup of plaque forms a necrotic core, much like the molten lava inside a volcano, a bubbling, boiling inferno, while the surface of the vessel threatens to crack under the pressure. At the site of the plaque, if you were to cut the blood vessel on the short axis, you'd see an eerie yellow smile as the plaque thickens underneath the inner round lining of the endothelium, shown in Figure 2.2.

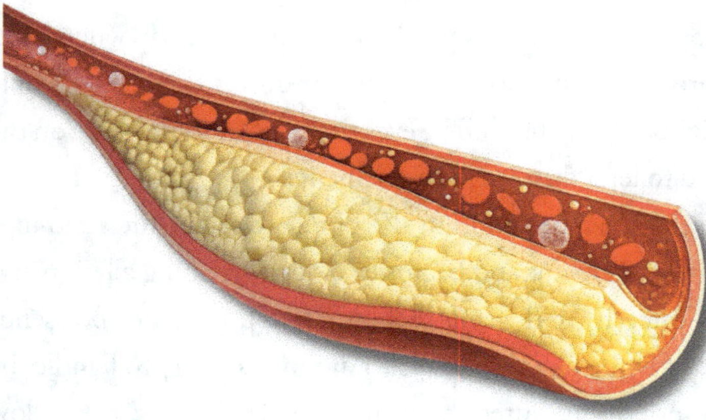

Figure 2.2

Calcification is the body's alternate response to plaque, assuming there aren't enough HDLs and LDL receptors in the liver to actually reverse it. Essentially, the body attempts to harden, or calcify, the plaque, much like hot lava cooling to stone. The same process also happens when you break a bone. As the inflammation at the site of the breakage heals, the body creates a calcified callus to protect the vulnerable bones as they regrow. At first, when the plaque hardens, it creates a large bump protruding into the lumen of the blood vessel—just like a bone callus over a healing fracture. But eventually, as it continues to harden and compact, the yellow brick road will smooth itself out and the lumen will restructure itself to restore its original diameter. In fact, the yellow brick road will actually be *stronger*, just as bones are often stronger after a break.

Cardiologists use a calcium score to determine whether plaque exists. A calcium score is one of the key tests we conduct as part of the Boone Protocol, which you'll learn more about in Chapter 9. Any score above zero isn't ideal—but, perhaps counterintuitively, if you *do* have a score above zero, then we want to continue growing that score as quickly as possible to cool the hot lava and prevent it from rupturing. This can be alarming for some patients, who may come to us with a calcium score of two hundred that increases to a thousand or more after treatment, but that's exactly what we want to see. It's a sign that the hot plaque is turning to stone. As you'll see in Chapter 8, the Boone Heart Protocol for addressing atherosclerosis rests on two strategies: one, increasing the number of apoB-clearing LDL receptors and apoA1-bearing HDL garbage trucks in the blood to clear away other apoB-bearing LDLs, and two, speeding up the healing calcification process to prevent further damage and reduce inflammation throughout the blood vessels. Just recently, I was looking at the results of one of our patient's CT angiograms. We could see the amount of calcified plaque in this patient doubling, meaning that the amount of hot lava in his veins was being cut in half as the body heals itself. I like to think of it as God making his own version of a stent.

Unfortunately, for most of us, this process doesn't happen fast enough. Left untreated, most plaque in our vessels is composed of only about 20 percent calcium; 80 percent is hot, soft, vulnerable plaque ready to erupt. It is this plaque that builds up in the blood vessels of unsuspecting victims like Jim Fixx, and which constitutes our greatest risk for dying of heart disease.

The Risk Factors

If dietary cholesterol alone isn't the cause, then what makes someone in prime physical condition, like Jim Fixx, who ate healthily and exercised every day, so vulnerable to atherosclerosis, while someone like my father, who feasted on Wendy's chili and chalupas for most of his life, might live to nearly a hundred? Unfortunately, the answer is a perfect storm of preconditions: genetics, high blood pressure, and lifestyle factors.

Genetics

Contrary to popular belief, the accumulation of LDLs within the lining of our blood vessels has less to do with the cholesterol-rich foods that we eat, and more to do with our genetic makeup, particularly if you have an extra sticky variant of lipoprotein, called lipoprotein-a, or Lp(a)—usually pronounced "L-P-little-a."

Clearly, LDL particles marked by ApoB are responsible for most atherosclerosis from a cholesterol perspective. However, the presence of Lp(a) is emerging as a major risk factor for atherosclerotic heart disease. It was discovered in 1963 by Norwegian geneticist Kare Berg as he was studying different types of lipoproteins in his early thirties. It proved to be such an important revelation that he devoted the rest of his life to studying this one specific lipoprotein, publishing hundreds of papers and several books on the topic. It was Berg who first realized the connection between Lp(a) and heart disease, noticing early on in his studies that it was a genetic trait passed down through families. Lp(a) can be found in 20 percent of the population and tends to be slightly more common among African Americans, although none of us should assume we lack this gene without specific genetic testing. In fact, many people, including most primary care physicians, ignore Lp(a) and neglect to test for it because it is so common. But the fact that it's common does not mean it's normal. Ideally, we should be checking patients for LP(a) right along with cholesterol and blood pressure as another indicator, if not the primary indicator, of the cardiovascular system's health.

Part of the reason LP(a) is not tested for is that it's a difficult problem to solve. Why bother testing for it if we can't do anything about it? LP(a) doesn't go away, even with good diet and exercise habits. But our knowledge and understanding of LP(a) is still growing, and although we may not be able to control its presence in our body, we can take extra precautions to regulate our lifestyle choices accordingly if we know we have it. LP(a) is particularly vulnerable to and exacerbated by smoking. My wife, three of my four children, and my son-in-law all have LP(a). While I was never able to run the test, we can reliably assume that my wife's father, Don Felgate (mentioned in the Introduction), also had

LP(a), and that, coupled with his smoking habit, was the cause of his untimely death in his fifties. Lp(a) is both atherogenic (plaque-causing) and thrombogenic (clot-causing). A daily aspirin may be especially beneficial in those with Lp(a). Had my wife's father, Don, known of his Lp(a), even a simple daily aspirin would have prolonged his life. Grandpa Don would be happy to know that his children and grandchildren will be protected by PCSK9 inhibitors (currently available medications which lower Lp(a) by 30 percent), as well as emerging drugs like Pelacarsen and Olpasiran, which promise to lower Lp(a) by up to 90 percent.

Another serious genetic cholesterol problem like Lp(a) is heterozygous familial hypercholesterolemia (HEFH), which is a multifactorial genetic mutation in the LDL-receptor gene—that is, the gene that provides our cells with instructions on how to make the LDL-receptors that allow cells to use and clear cholesterol. These receptors were discovered by Nobel-Prize-winning doctors Michael Brown and Joseph Goldstein in 1973. Brown and Goldstein found that all cells have LDL receptors, which enable the cell to grab the LDL and use the cholesterol for its structure, or for the production of hormones or digestion. But people with HEFH have a mutation with these genes and therefore have fewer LDL receptors than normal, and sometimes no LDL receptors at all (homozygous familial hypercholesteremia (HoFH)). Brown and Goldstein's discovery of these receptors was a significant advance in our understanding of the process of atherosclerosis, winning them a Nobel Prize in 1985.

As with LP(a), without enough of these LDL-receptors, the unused LDL will flood the bloodstream and become embedded in the endothelium. HEFH can be found in one in five hundred people, making it surprisingly common—and yet, again, few doctors test for it. It is also especially deadly, causing accelerated atherosclerosis in most patients who have it before age fifty. Those with HEFH can even die as babies. We sometimes find evidence of HEFH in the children we test through the EveryHeart Foundation. A student athlete might only be twelve or thirteen, but if we find HEFH, their arterial age may already be as high as thirty to fifty, meaning the lining of their arteries is already thick with plaque and calcification. In other words, in people with HEFH, the process of atherosclerosis occurs three to five times faster than in patients without HEFH. The risk for myocardial infarctions (heart attacks) in these patients is incredibly high. Unlike LP(a), however, HEFH is highly treatable with a variety of statins and cholesterol inhibitors that can lower the overall LDL concentration in the blood. There's simply no reason why patients with HEFH need to ever experience a life-threatening cardiac event.

Let's face it, though: most people in the industrialized world have too many atherogenic cholesterol particles and too few receptors to clear the excess. The lessons in this book apply to all of us, not just those with rare genetic mutations. When we know about our risk factors for the deadliest disease in human history, we can take action to avoid the triggers and mediators that would launch and worsen the disease. Thankfully, as you'll see in Chapter 8, there's much that can be done, especially with innovative medicines, to render even potentially lethal genetic traits like LP(a) and HEFH completely harmless.

Unfortunately, most people never know if they have LP(a) or HEFH. At a routine physical, or perhaps in response to some mild chest pain, most doctors will perform a standard cholesterol check or a treadmill stress test, if they're really concerned. But waiting for chest pain or stress test abnormalities can be fatal. Usually, by that point, the plaque inside of us has already been growing for twenty to forty years. If the stress test shows any abnormalities, an angiogram may be recommended to check for blockages throughout the blood vessels. Using special dye to illuminate the blood flow throughout the body, the angiogram will determine whether any blockages *already exist*. This can be lifesaving in terms of preventing that blockage (or clot) from blocking blood flow to the heart or the brain and causing a heart attack or stroke, as we'll discuss in the next chapter. But the angiogram is effective for diagnosing serious blockages or ruptures once they've already happened; it's impossible to tell from an angiogram whether plaque is accumulating within the lining of the blood vessels, just waiting for the right opportunity to burst through. Thankfully, the procedure to identify existing plaque within the arteries is even less invasive and expensive than an angiogram; a simple ultrasound of a few key arteries and a coronary calcium score can help us determine whether plaque is building up. By assessing the most likely places where plaque tends to show up (remember, it will usually accumulate in areas where the flow changes or where blood pressure is low), we can assume the rest of the highway is similarly clear or at risk.

The danger, though, is that most people never get checked until they start showing symptoms—and by then, it's already too late. The dominoes are set; all that's needed is the perfect catalyst to cause a chain reaction, resulting in a heart attack or stroke. As we've already begun to see, even those who live a fairly healthy lifestyle, with adequate nutrition and regular exercise, may still be at increasing risk for atherosclerosis as they age.

High Blood Pressure

Another risk factor for atherosclerosis is high blood pressure, or hypertension. Blood pressure is a measure of the force exerted as your heart pumps blood throughout the body. It is reported in two numbers, usually seen stacked on top of each other. The top number is your systolic blood pressure, or the force when there's a surge of blood as the ventricles in your heart contract. A healthy resting systolic blood pressure is less than 120, although this number is quite variable and rises during physical or mental stress. If we see a consistently elevated systolic blood pressure, then the blood vessels have to work harder to regulate the flow, like trying to control a firehose with the water turned on full volume. The bottom number is your diastolic blood pressure, or the force between beats when your heart is resting. Usually, we want to see this number below 80. If both numbers are elevated, we know that the blood vessels are constantly working overtime to control the pressure. Eventually, they will weaken and threaten to burst. The heart muscle is already damaged, as well, as it adapts to this high-pressure system.

Like atherosclerosis itself, we are often predisposed to high blood pressure through our genetics. That is, if your relatives—not just your parents, but also your grandparents, had high blood pressure, you're likely to have it as well. One of the many findings from the continued Framingham Heart Study has been that high blood pressure is heritable through at least three generations.

High blood pressure is also a result of environmental factors. A 2023 study of the effects of noise and air pollution on over 240,000 participants led by Jing Huang at the University of Beijing found that long-term exposure to noise pollution (such as sirens and other traffic noise) leads to an increased heart rate and blood pressure. The impact is even worse when air pollution is present as well. Blood pressure also rises in cold temperatures and at high altitudes, such as when flying or hiking to the top of fourteen-thousand-foot-tall mountains, as many unfortunate Coloradans have discovered.

There is a circadian rhythm to blood pressure, which tends to surge in the morning, peak in late afternoon, and then decrease in the evening and into the night. This is why so many heart attacks occur in the morning, as Mark's did. The anticipatory stress of facing the workday also exacerbates this phenomenon.

Lifestyle

While genetics undoubtedly has the greatest impact on atherosclerosis, our lifestyle choices do play a major role in its development as well.

Certainly, if we're flooding our bodies with cholesterol-rich foods, we may tip the scales in the balance between HDLs and LDLs, adding to our overall lifetime cholesterol burden. The good news is that it's unlikely that diet alone would be enough to cause severe, early atherosclerosis; lifestyle is more of a risk if you have the LP(a) or HEFH gene. That's not to say you should eat whatever you want, as there will undoubtedly be other consequences—but for the purposes of atherosclerosis, the cholesterol found in our food is unlikely to make a significant difference, although every preventive cardiology effort begins with the foundation of a heart-healthy diet. Smoking, however, is a critical risk factor for atherosclerosis, as it lights all the fiery furnaces of harm in our body, particularly if you have the LP(a) gene. Sleep deprivation, especially the disturbance of our circadian rhythm and the resulting disruption of our natural blood pressure patterns, can also weaken the blood vessels and predispose us for a critical heart event.

Stress

The final and most significant lifestyle risk for atherosclerosis is stress, both acute and chronic. My own awareness of the impact of stress on heart disease began after my residency. I had my eye on the Cooper Clinic, working under the renowned physician and the "founder of aerobics" Dr. Kenneth Cooper.

I had first met Dr. Cooper in 1975, when I was the local fitness expert at the YMCA in Des Moines, Iowa. I was in charge of our annual jogging fundraiser, encouraging people in the community to jog a certain number of miles to get prize T-shirts, which were awarded at a large banquet we held every year. I knew about Dr. Cooper's reputation in aerobics and his pioneering of treadmill testing, which most cardiologists thought was a waste of time. Naturally, I thought there couldn't be anyone better to come speak at our banquet, so I reached out and asked Cooper to come. He agreed and his team turned the event into a tour of Des Moines to promote his latest book. Cooper was a hit at the banquet, and I had the pleasure of being his chauffeur for a few days.

At the time, I was still trying to decide what to do next with my career, and on one of our rides I asked him if he thought I should get a PhD in exercise physiology. He advised against it, just as my wife had.

"If you're a PhD, you can't prescribe medicine, and you can't take care of heart attacks in the hospital," he said. "In all of my world, the most powerful people have MDs, because they don't need anybody's permission to do anything." I told him that my now-wife Colette had also suggested the idea of getting an MD, and Cooper's encouragement further fostered that idea.

Eight years later, after my residency in Portland was complete, I tried to get a job with Cooper in Dallas, where a large estate with a mansion had been donated to further his research and allow him to continue treadmill testing and further his pioneering work in preventive medicine. But the timing didn't align. They didn't have any openings for me. Instead, I heard about a program in Phoenix, Arizona, spearheaded by cardiologist Robert S. Eliot.

In the early eighties, Eliot and his team had used the phrase "hot reactors" to describe certain physiological traits of people who are vulnerable to heart attacks. In 1983, a profile of Eliot in *The New York Times* described the idea this way:

"Hot reactors tend to overrespond physiologically under stress, although this overreaction is not apparent to anyone, including the hot reactors themselves. Their blood pressure rises sharply and other circulatory system responses may also occur: The output of the heart may increase or decrease, and the blood vessels may become more or less resistant to the flow of blood."

I recognized that Eliot was trying to objectify, quantify, and make physiologic the response of stress. People like you and me and many of my patients who work hard don't necessarily have nervous breakdowns, but that doesn't mean the stress we experience daily doesn't have an impact. Our bodies pay us back through everything from migraine headaches to ulcers to chest pain to autoimmune diseases, and especially atherosclerosis. Eliot himself had had a heart attack, which he believed had been caused by stress. And so Eliot began pulling out the anecdotes of stress from his patients and finding ways to systematically measure their experiences. As soon as I heard about it, I knew that I wanted to be part of this journey. I called Eliot and asked him if he had any openings. He didn't.

"Well, I'd like to come down and talk to you anyway," I urged.

"No, don't even bother," he answered.

I went down nonetheless, because I knew they'd never find anybody like me. This isn't to say I possessed any unique talent, but rather that my passion, commitment, and focus were rare among physicians at the time. I always tell people that if you find a place where you really think you belong, you can make your own job because they cannot resist you if you're good enough. And that's what happened.

I showed up one morning at Eliot's offices and he palmed me off on another cardiologist on his team, Hugo Morales-Ballejo, an Argentinian doctor, who explained the work they were doing. They were beginning to assess for hot reactors with innovative tests, like having patients put their hands in ice water or say the alphabet backwards or do mental mathematics while measuring their blood pressure and heart rate. The results were astounding, but even more exciting was the implication that we could find the hot reactors of the world and save them from potentially devastating cardiac events. Hugo Morales-Ballejo told Eliot they had to hire me, and after spending that day together in 1986, we became lifelong friends.

I stayed with Eliot in Phoenix for the next year, helping him and his team develop and trial these new tests. We worked with hospitals who were intrigued by our methods and wanted us to offer preventive services to high-end customers as a way to increase their revenue. Of course, what they didn't realize was that they were actually decreasing their revenue in the long run because we were able to help prevent cardiac events for these patients. Every now and then we would discover somebody who needed a valve replacement, but most of the time, we would just facilitate an even more aggressive preventive and stress medicine regime. After a year, Eliot moved his practice to Denver because another hospital was interested in the program I now directed. A few years later, Eliot's contract in Denver was finished, and I decided to continue the work on my own, offering many of the same tests to interested patients at a tenth the price we had charged through the hospital. I took the lessons about stress testing and the potential risks for hot reactors with me as I began my new venture, expanding the horizons of preventive cardiology.

At the time, I knew that atherosclerosis, the slow, silent buildup of plaque within our blood vessels, was just the preparation for the main event. As Glagov discovered, for most of the time that plaque is accumulating, the lumen remains clear until that necrotic core reaches the critical threshold of about 40 percent. At that point, the weakened endothelium can no longer hold back the plaque. The wall of the cream filled donut is full up. The moment there's an increase in blood pressure, such as Mark experienced on his stressful morning, or as Jim Fixx had on his hot afternoon jog, the volcano erupts. But the impact of stress is so complex and our understanding of it has progressed so much in recent decades that we'll devote the entirety of Chapter 6 to it.

When I chose preventive cardiology, I had no idea the wild frontier I was stepping into. Wandering through the uncharted territories of this silent, deadly disease would prove to require a level of persistence and tenacity to rival that of my ancestor Daniel Boone (Yes – that Daniel Boone).

CHAPTER 3

The Clot That Kills: Thrombosis

Family legend, and good deal of sleuthing, has it that I am descended from the famous frontiersman Daniel Boone. Growing up, all the Boone children knew that my father, Stewart, was a descendant of Daniel's brother, Samuel. In adulthood, my more genealogically inclined relatives did the research to prove the family story was true. So, while I personally can barely start a fire with a lighter, I sometimes think that I may have inherited something of Daniel Boone's pioneering spirit: an impulse to explore new worlds or tackle unresolved problems. But beyond serving as a fun piece of family trivia, I find that my great-great-great uncle Daniel provides a helpful illustration of the counterintuitive way that atherosclerotic heart disease kills us.

Picture the famed frontiersman carving a path through the bush in the Cumberland Gap on a cool morning, clad in a deerskin jacket and his signature coonskin cap. The untamed wilderness is dangerous, so he carries a musket and wears a long knife dangling from his belt. They say that In those days, the forest was so dense that you could move through the treetops from the East coast to the Great Plains without touching the ground. The sun is rising, but visibility is still limited. Daniel is alone on this morning, far from his camp, scouting out a new westward route through the Appalachians. Perhaps he can trap a beaver or find a deer for dinner while he's at it.

Daniel emerges from the deep woods into a small clearing where, much to his surprise, he finds a small black bear. Equally surprised, the cub huffs in alarm, pacing frantically, signaling for its mother to protect it from the unexpected intruder.

Immediately on high alert, Daniel looks around, knowing that the mother must be nearby. He hears a rustle to his left, and suddenly, he spots her large black pelt as she charges through the brush. He stumbles backward, raising his musket, but the bear is too quick. It closes the distance in the blink of an eye and swipes at Daniel's shoulder, claws piercing deep into flesh. He is knocked off his feet, his shoulder bursting with pain. He knows that most black bears don't want a confrontation; already, the mother has gathered her cub behind her and turned to survey his next move. She shakes her head and lets out an explosive blow through her nose, pawing the ground anxiously. A warning. He raises his rifle and fires a warning shot in return. It has the desired effect; she has no wish to fight. The bears turn in haste and disappear into the shadows.

Daniel slumps to the ground. The leather covering his shoulder is soaked with blood. His vision is blurring. Shock has dulled the pain, but Daniel feels weak and vulnerable. Luckily, his body is already working to save him, having developed a response to these types of traumas over millions of years. In Daniel's bloodstream, tiny particles called platelets are rushing to the site of the injury. As platelets gather, a clot begins to form, slowing and then stopping blood loss. In the coming days, a protective scab will form over the wound to prevent further bleeding and infections.

No stranger to quick thinking in times of crisis, Daniel stuffs a handkerchief under his shirt, wincing in pain as he presses it firmly to his wound. He pulls his jacket tight on top of the wound to secure the makeshift bandage. His shoulder hurts. His right arm is near useless. He isn't able to reload his musket with one hand. But he has his knife and his senses, and he hasn't passed out from blood loss. Daniel takes a pull off his waterskin, then summons his strength and stands. After a moment to steady himself, he starts the seven-mile hike back to camp.

In this fictional scenario, Daniel is able to walk away from a potentially deadly encounter due to a combination of quick thinking, experience, and, most crucially, his body's natural defenses. But the experiences of pioneers like my ancestor provide a compelling, outwardly visible illustration of the internal processes happening within us. When we are injured, the chemical responses to trauma—adrenaline, shock, clotting—save our lives and allow us to keep hunting, farming, fighting, building, and progressing as a species. What's less visible to the naked eye is the way that our bodies, particularly our arteries, heart, and veins, do the same thing inside us.

Thrombosis is the body's process of creating clots, originally meant to heal injury and protect us from further harm. After decades of slowly building up plaque, a stressful morning or a hot afternoon jog can be all the catalyst our weakened endothelium needs to rupture. When the volcano erupts, bursting through the fragile lining of our arteries, the body responds just as it would to a swipe on the shoulder from a bear's claw. Platelets rush to the scene to stanch the outpouring of harmful plaque into the bloodstream, forming a clot—or thrombus—just like Daniel Boone stuffing his handkerchief under his shirt.

Unfortunately, however, this adaptive response to injury that used to save us when predators were the most likely cause of death is the very thing that now threatens our lives. It is lifesaving when we're attacked by bears or tumble off a mountain bike, but it becomes lethal when it happens inside our blood vessels as a result of atherosclerosis. The blood clot bogs down traffic in our superhighways like a major accident; blood can no longer make its way to the heart muscle with the nutrients and oxygen we need. If a clot prevents blood from reaching the heart, it triggers a heart attack (myocardial infarction); if it lodges near the lungs, it will cause a pulmonary embolism; and if it is near the brain, it causes a stroke. All of these often-deadly events result from the same biological process that used to save us. A ruptured plaque inside a coronary artery causes the same response as Daniel's bear claw wound, there's just no room for the thrombosis (clot) in a one-millimeter tubular coronary artery. The same thrombosis that gave life in the late 1700s kills in the early 2000s.

This problem is particularly acute for those who carry the gene for Lp(a). It's likely that so many people evolved to have this gene for extra sticky cholesterol precisely because of its ability to help us quickly form lifesaving clots. Now, as our lifestyles have changed and our biggest predators are not outside of us but within us, LP(a) has increased its carriers' risk for dying early of heart disease. As with many life-threatening issues within the body, the protective mechanisms that humans evolved to keep us safe haven't kept pace with the evolution of our society, and so our bodies' reactions are several thousand years behind the times. As Dr. Nadine Burke Harris writes in *The Deepest Well*, "The difference between adaptive and maladaptive reactions is all about the when." Those few who seek dangerous adventure today will still be protected if they are injured, but their coronaries are more likely to kill them than the bear.

The clot that used to save us has become the clot that kills.

How Clots Form

Imagine a volcano erupting for a moment. I always think of Mount St. Helens, which we could clearly see from our home in Portland, Oregon, during my residency, and which is the most active volcano in the continental United States. The last time Mount St. Helens erupted was in May, 1980, just a few years before we moved to Portland. That day, the first vicious explosion sent a stream of lava, ash, and volcanic steam over fifteen miles high into the atmosphere at more than 300 miles per hour for fifteen minutes. The blast cloud stretched over seventeen miles to the north and covered more than 200 square miles, causing a massive landslide, the largest in recorded history. The eruption and resulting landslide killed fifty-seven people in its wake. When we lived there, Mount St. Helens loomed large and menacing indeed in everyone's minds.

When the hot lava plaque lying underneath the surface of the endothelium ruptures, the platelets floating around within our blood sense the break and begin to attach themselves to the vessel at the site of the injury, essentially trying to plug up the top of the volcano. The platelets, which Siddartha Mukherjee calls "the healing cells," then attract more platelets from the blood, linking together to form a long, thick protective barrier, spreading lengthwise throughout the blood vessel as more and more platelets join in, as shown in Figure 3.1.

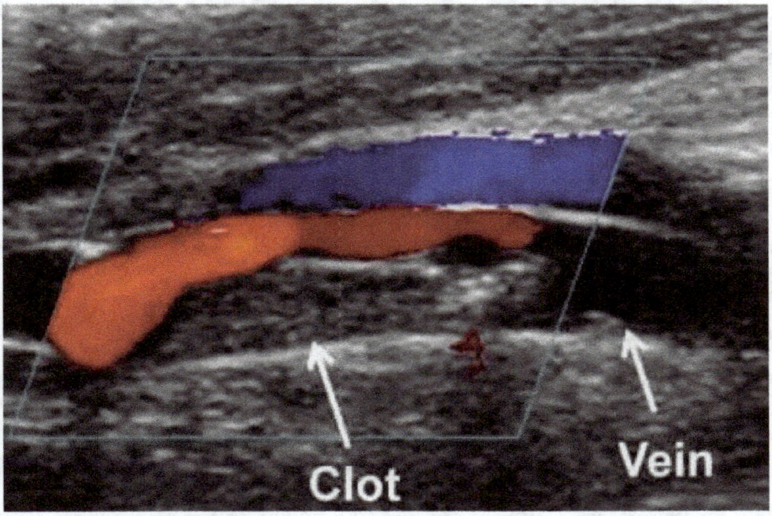

Figure 3.1

The platelets also attract a protein called fibrin, which is responsible not only for stopping the flow of blood (vital with open wounds) but also eventually shrinking the clot and resulting scab to restore blood flow to the damaged area—both functions that become critically important in the progression of heart disease. Together, the fibrin and the platelets create a dense, protective coagulation to stop the wound and to begin the healing process. The plaque that shot forth from the rupture will be cleared away along with all the other bad product in our bodies; the lava will cool on the surface of the yellow brick road, and blood flow will be restored.

It may surprise you to learn that this process likely happens thousands of times when we have the right balance of clot-forming and clot-clearing cells within the body. Anyone who has a coronary calcium score above zero has had clots come and go, healing the plaque and leaving behind a calcium-laden matrix. Most of the time, clots simply go away on their own, leaving a calcified scar behind. The right lifestyle choices (like not smoking and eating healthily) and the right medications can hasten their healing. But for those of us with an abundance of atherosclerotic plaque, certain genetic factors (which we'll discuss later in this chapter), and conditions or lifestyles that exacerbate the presence of clot-forming cells, the clot will continue to thicken on top of the ruptured plaque. The platelets sense that they may have the damage under control, but they have often blocked the lumen in their quest to heal the rupture and, in fact, have caused a heart attack or stroke, all with the best intentions.

Now, when a volcano like Mount St. Helens erupts, the explosion can shoot up as far as the force of the blast will take it—even fifteen miles into the atmosphere. But when the hot plaque in our arteries erupts, there *is* an upper limit, a much smaller height which the blast can reach—and so instead, it spreads out. The roof above these tiny blasts is measured in microns, not miles. Remember the 1998 movie *The Truman Show*? In the movie, Jim Carrey's character learns that he's been living inside a giant production studio; his entire life has actually been filmed and put on display for viewers in the real world to see. Our arteries are kind of like that production studio; the sky is actually a low ceiling, so when the plaque erupts, there's nowhere for it or the resulting clot to go. As the clot develops to cover and to plug the rupture, it stretches out upward and lengthwise within the long tube of the blood vessel, blocking the blood flow completely in many cases. At this point, the body may try to reverse its clotting activity because of the reduction of blood flow downstream. In fact, if you make it to the ER, they may give you clot-busting IV medicines. Indeed, we recommend chewing four baby (81 mg) aspirin as soon as chest pain from reduced coronary artery blood flow is noted to start this "clot-busting" process. It's like stopping a train at full throttle and putting it in reverse.

The problem is that our bodies are very good at forming clots—too good, in fact. In *The Song of the Cell*, Dr. Mukherjee writes about how this lifesaving reaction has turned into a life-threatening one: "If the vagaries of ancient human life involved plugging wounds to maintain homeostasis, the vagaries of modern life have unleashed the opposite problem: *too much* platelet activation."

But to continue our work to prevent the overzealous action of platelets and fibrin to form clots, we need to better understand the three conditions that predispose us to clots in the first place.

The Deadly Trifecta: Virchow's Triad

Rudolf Virchow, a Prussian doctor, political activist, and researcher in the nineteenth century, is now considered one of the pioneers of cellular biology. Virchow was fascinated by the pathology of disease—that is, figuring out why and how diseases begin and develop at the cellular level. He is the first to use the term *thrombus*, from the Greek meaning a "curdling" or "clumping," in his revolutionary 1856 book *Cellular Pathology*, to describe the masses that doctors sometimes found inside veins and arteries. Popular doctors of his time believed that these masses were made of pus caused by inflammation, but Virchow showed that they were actually coagulations of blood (now, more specifically, we know that they are coagulations of platelets and fibrin), formed through a process which he called *thrombosis*.

Virchow still wanted to understand where these coagulations came from—in particular, how they end up in various veins and arteries throughout the body. There are three essential conditions that predispose us to the formation of thrombi: damage to the lining of the blood vessel, hypercoagulability, and stasis. We call these conditions "Virchow's triad" because he correctly identified their presence and involvement in the clotting process, even though, as you'll see later on, Virchow himself had no idea that these conditions actually *cause* blood clots. Like many scientists, he was so close to the answer he was looking for, and it is only in the 150 years since then that we have come to understand how these three factors come together to form a trifecta of conditions for disaster.

The first requirement, damage to the intima media, we've already discussed. This comes primarily as a result of atherosclerosis, as the LDL particles infiltrate the endothelium and create an inflammatory response inside, forming that necrotic core of plaque. When it ruptures, the damage necessary to begin attracting the platelets is done. However, damage to blood vessels can also be a result of physical trauma, intravenous drug use, or surgery. Those with cancer, especially those undergoing treatments for cancer, are at higher risk for blood clots because of the increased inflammation throughout the body and the damage to the endothelium that occurs during radiation and chemo.

The second requirement, hypercoagulability, is an elevated tendency of blood to form clots. Hypercoagulability happens as a natural result of aging, but it can also be caused by inflammation, pregnancy, hormone therapies (especially for women taking estrogen), obesity, diabetes, cancer, and genetics (including Lp(a), but also others, as we'll discuss in a moment). In pregnant women, levels of fibrinogen, the pre-clot form of fibrin, naturally increase throughout the pregnancy as the body prepares for labor, to prevent blood loss for the mother during birth. Without this increase in fibrinogen, women are likely to have hypofibrinogenemia, a disorder that can cause recurrent miscarriages and severe hemorrhaging during birth. The downside of this increase in fibrinogen, however, is that women are at an increased risk of developing blood clots, which is why obstetricians often recommend their patients take baby aspirin throughout pregnancy.

Patients with obesity and diabetes also show increased levels of thrombin, the enzyme that converts fibrinogen to fibrin, and other coagulating blood factors.

In total, the blood holds thirteen factors that contribute to the clotting process, typically numbered with Roman numerals. Fibrinogen (Factor I) and prothrombin (Factor II, the precursor to thrombin) are just two of these factors. Besides pregnancy, obesity, and certain diseases, genetic abnormalities can also cause increased levels of these clotting factors. The most common are Factor II prothrombin mutation and Factor V Leiden. With a Factor II prothrombin mutation, the level of prothrombin is found to be elevated in the blood. Factor V is the factor responsible for shutting off the formation of a clot once the platelets have the damage to the blood vessel under control; a mutation with this gene (Factor V Leiden) means that the clot will continue forming beyond what's necessary. Both mutations significantly increase the risk of clots, particularly of deep vein thrombosis and pulmonary embolisms, which we'll discuss in a minute. Factor V Leiden is prevalent in about 1 to 5 percent of the general population, but in those with venous thromboembolism, a condition where blood clots form in veins, the prevalence is between 10 and 20 percent.

Another genetic precursor to thrombosis is the presence of the apoliprotein-e4 (ApoE4) gene marker. Everyone has the apolipoprotein-e gene, but it comes in four different variants, or alleles. A study of the progression of early subclinical atherosclerosis (PESA) screened over four thousand middle-aged participants for evidence of cardiovascular disease, finding that carriers of the ApoE2 gene have the lowest risk of CVD; ApoE3 carriers had a slightly increased risk; and ApoE4 carriers have a significantly higher risk of developing plaque in the lining of their arteries, as shown in Figure 3.2.

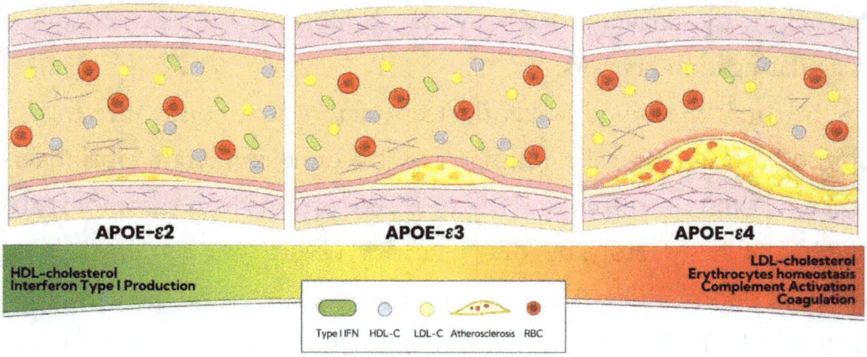

Figure 3.2

The ApoE4 variant causes an increase in fibrinogen (Factor I) and promotes inflammation, which, again, was probably a *good* thing in millennia past. Now, however, this gene is an indicator of a higher likelihood of developing Alzheimer's and dementia. In 2022, while filming the docuseries *Limitless*, actor Chris Hemsworth discovered that he carries two copies of the Apoe4 allele, one inherited from each of his parents, essentially doubling his risk of acquiring Alzheimer's later in life. He was offered the option of not revealing this finding on the show, but ultimately decided to make the information public in an effort to raise awareness of the risks, which was the whole point of the docuseries to begin with. In an interview with *Vanity Fair*, he noted that he was grateful to be made more aware: "Like everything in the show I went, 'Okay, great. I now have to work on this more.' If you look at Alzheimer's prevention, the benefit of preventative steps is that it affects the rest of your life. When you have a predisposition to cardiovascular heart disease, cancer, anything—it's all about sleep management, stress management, nutrition, movement, fitness." It's a little more complicated than that, but as you'll see, he's not far off. After all, a great poker player can be dealt a bad hand and still win!

The third element in Virchow's triad is statis—literally, a lack of movement. Bed rest during pregnancy, recovering from a surgery, or other periods of inactivity, such as a long flight, can all trigger the formation of thrombi. Thankfully, stasis alone is not enough to cause clots, otherwise we might all acquire them every night while we sleep! Rather, it's a lack of movement along with the other two elements in Virchow's triad that predisposes us.

Embolisms

In 2000, Kansas City Chiefs linebacker, Derrick Thomas, died after surgery for a spinal cord injury. The surgery was successful, but in recovery, Thomas was being moved from his bed to a wheelchair when a piece of a clot dislodged from wherever it had started and traveled up his bloodstream, where it was caught right in between his lungs and his heart. Blood flow to these vital organs was almost immediately cut off. All the preconditions for the clot had been met: injury to his blood vessels, hypercoagulation as his body tried to heal his other injuries, and stasis as he lay in bed recovering. Unfortunately, efforts to treat and resuscitate Thomas were unsuccessful.

In addition to helping us understand how we are predisposed to clots, Virchow was the first to realize that many thrombi are not isolated coagulations of blood unto themselves; they are actually break-offs of a larger thrombus elsewhere in the bloodstream. He named these satellite clots *embolia*, from the Greek word *embolos*, meaning a "peg" or "stopper." Virchow noted that clots can start as thin as "a knitting-needle," but may become prolonged and eventually reach the "thickness of a thumb," especially in the veins. When that happens, an increase in blood pressure, or the fibrin's natural tendency to break down clots, will cause a piece of it to break off. Ironically, Virchow himself didn't realize that the three conditions that we now know as "Virchow's triad" were precursors to all clots; he identified them as the *consequences* of clots and the precursors to embolisms specifically, which is accurate. In other words, the same three factors that cause clots (damage to the endothelium, hypercoagulation, and stasis) also cause those clots to break off into further embolisms.

It's when we move again after a long period of stasis, such as bed rest or recovery after a surgery, that clots in our veins are likely to be broken down into smaller pieces by the sudden resumed pressure of blood flow. The pressure causes a large chunk of the clot to break off, and then it is often carried along by the bloodstream until it becomes lodged in larger and more critical blood vessels, such as the arteries, causing more clots elsewhere, particularly the lungs. Virchow found in his studies that "Usually a fragment of the kind sticks fast where a division of the vessel takes place, because the diverging vessels are too small to admit it." Again, we see that the most vulnerable areas are places where blood flow is low or changes.

Thrombosis can occur throughout the entire body, not only as a result of atherosclerosis, but also due to any changes in pressure in the vessels. We saw this recently with one of my patients who has a genetic tendency towards clotting. After a long flight, he came in for a heart scan, and in the margins of that image, we noted a number of clots in his lungs. Remember, areas of low or changing pressure throughout the body are particularly vulnerable to not just plaque buildup, but also vessel breakage and clots. Like this patient, you may know people who tend to get blood clots on long flights, usually within the deep veins in their legs—called deep vein thrombosis—because of the change in blood pressure in the leg veins after prolonged sitting. These clots are insidious. They are notorious for causing embolisms that can slide out of your legs and travel up to the lungs, where they may result in a pulmonary embolism. Usually, this means near-instant death. Thankfully, in my world-travelling patient's case, we caught them in time and with the right medications were able to safely break up those clots and prevent them from becoming a much larger problem. To this day, that patient never stays sitting motionless for prolonged periods of time and takes daily aspirin, especially prior to international flights. As always, it is imperative to identify and be vigilant about your own abnormalities and unique risk factors in order to prevent disaster.

Varicose veins, again in the legs, also pose a greater risk for thrombosis and pulmonary embolism. In the long veins within the legs, a series of valves is responsible for pumping blood back up to your heart for reoxygenation, a critically important function, as they are working against gravity. When one valve breaks, as sometimes happens if you have chronically high blood pressure or other cardiac risks, blood accumulates at the next valve, putting greater pressure on it to send all of that blood back up to the heart, as shown in Figure 3.2. If the fragile veins continue to break under the pressure, a clot could develop in the growing static pools above the progressively more damaged valves of the leg veins. The low flow and pooled blood places a person at high risk of a pulmonary embolism.

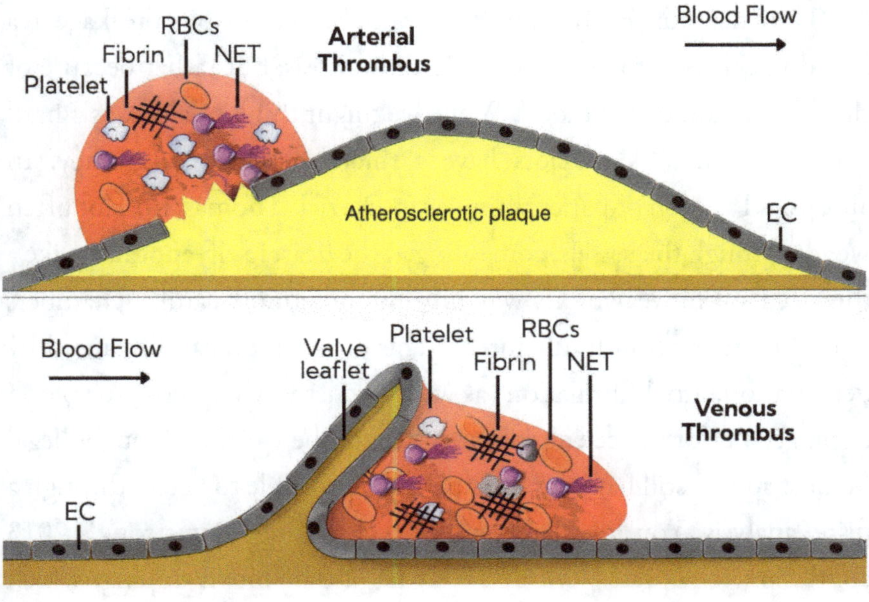

Figure 3.2 (https://www.ahajournals.org/doi/full/10.1161/ATVBAHA.119.312463)

In the image above:
EC = endothelial cell
RBCs = red blood cells
NET = Neutrophil Extracellular Trap (aka White Blood Cells Engulfing Cholesterol Particles that have entered the wall of the blood vessels)

Certain surgeries increase the risk of clotting: pelvic surgery in women, prostate surgery for men, and orthopedic procedures such as knee or hip replacements. In these instances, circulation is already inflamed. If a patient then is exposed to high altitude, or any other change in stasis pressure, their chance of a blood clot skyrockets. This happens most commonly in veins, which are like small streams carrying blood back to the heart for oxygenation, rather than arteries, which are like large rivers pushing blood out from the heart toward the organs. If a person also has Lp(a) or another genetic cause of hypercoagulability, their risk is even greater.

The final cause of thrombosis may not be because of a breakage in a blood vessel, as with atherosclerotic thrombosis, but rather because of the coagulation of the blood. While we might think of a clot as a hard scab, in reality, blood clots have a thick, jelly-like consistency. In autopsies I've seen of patients similar to Derrick Thomas, the clot often wends through the vessels, congealing in the heart itself, appearing like a thick red snake weaving its way into the innermost cardiac chambers where it strikes with deadly intent. This coagulation can be caused by dehydration, atrial fibrillation (as we'll discuss in Chapter 4), or blood doping, a performance enhancement technique which, although illegal in most sports, still happens because it is impossible to detect through a urine analysis. You might remember cyclist Lance Armstrong's 2013 confession to Oprah Winfrey about his use of blood doping, which helped him win seven Tours de France. Now we know that blood doping significantly increases the risk of blood clots, and hopefully more athletes will think twice before resorting to such measures.

The Final Blow: Heart Attacks, Strokes, and Pulmonary Embolisms

The clot most likely to kill us is an arterial thrombus, a clot inside a major artery, especially the aorta. Our arteries carry blood away from our hearts and out to all our other organs, but it is through the coronary arteries, offshoots from the aorta, that oxygen-rich blood flows to feed the heart itself. If the entire lumen, or the open center, of the coronary arteries becomes blocked, the flow of blood to the heart stops and the muscle is suddenly starved for oxygen and nutrients (Figure 3.3). As Mark experienced in Chapter 1, a tight, squeezing pain called angina spreads throughout the chest, as if the heart is being choked, which, in effect, it is. If blood flow is not restored quickly, the muscle will start to die. This happens faster than you might expect; often, even lifesaving measures like CPR or defibrillators come too late. The damage is swift and merciless. If a heart attack is detected quickly enough, a patient may be able to chew five or six aspirin to try to quickly break up the clot. After decades of lying seemingly dormant—a quiet, clear lumen hiding the slow buildup of LDL particles in the endothelium, the oxidation of those particles, and the formation of a necrotic core of smoldering hot plaque—the blood clot and subsequent heart attack occurs with surprising, devastating fierceness once the volcano erupts.

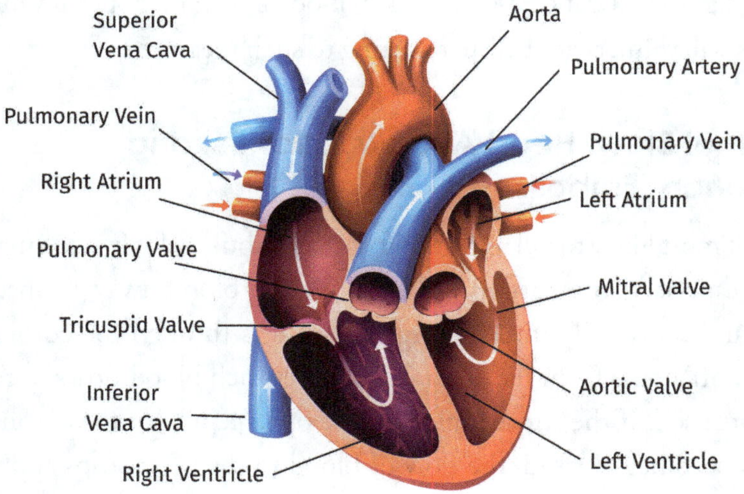

Figure 3.3

Myocardial infarctions, or heart attacks, are the most common cardiac events. It's estimated that about 805,000 new or recurrent heart attacks happen every year, over 109,000 of which are fatal. These statistics are even more shocking when you consider that every forty seconds, someone has a myocardial infarction. Thankfully, although the incidences of heart attacks have greatly increased in the last century, so has awareness around the signs and recommended responses. Nearly everyone knows about the tell-tale angina, or chest pain, that precipitates a myocardial infarction, although angina can also manifest as pain in the left arm, shoulders, neck, jaw, abdomen, or back. We know that chewing aspirin can help to break up a clot. Automated external defibrillators (AEDs) can be found hanging subtly on the walls in most public spaces, and many people are CPR-certified. According to the American Heart Association, CPR can double or triple the chances of survival, if performed immediately. The power of this lifesaving knowledge cannot be overstated; the better we get at recognizing the signs, and the more people who know how to respond appropriately, the more we'll see the fatalities from myocardial infarctions decline. Still, we have a long way to go. In a 2015 survey, only 18 percent of responders reported being up to date on their CPR certification.

Stroke is another leading cause of death in the United States, with over 795,000 people suffering from strokes every year, over 162,000 of which are fatal, according to the CDC. Further, about 87 percent of the total strokes that occur are ischemic strokes caused by atherosclerosis. In an ischemic stroke, a clot or embolism lodges in the branches of the carotid arteries, blocking the supply of oxygen to the brain and causing immediate damage to the brain tissue surrounding the clot.

Several years ago, a surgeon, who coincidentally lived down the street from us, was out running a marathon in San Diego. At about sixty years old, our neighbor appeared to be in prime condition. He was fit and healthy, running marathons regularly. This time, in the middle of his race, he suffered from a massive stroke. Our neighbor never ran again. A few years later, his doctors discontinued life support and he died. I've always wondered: If we could have seen inside his arteries prior to that marathon, could we have found the atherosclerosis that led to the blood clots that eventually killed him? If we had been able to do genetic testing, would we have seen that he had Lp(a) and advised him to take a daily aspirin, or perhaps avoid the marathons altogether? Armed with this knowledge, would he have known to take extra precautions against dehydration and other risks?

These are the questions I wonder about when anyone suffers from a highly preventable, devastating event such as a heart attack or stroke—the sudden end of the line after the slow and subtle progression of atherosclerosis.

Thankfully, most people who suffer from strokes don't endure the same fate. As with heart attacks, we've gotten better at noticing the symptoms and responding quickly. The National Stroke Association has created a helpful acronym (FAST) to help bystanders recognize the signs of stroke:

- F: Face drooping

- A: Arm weakness
- S: Speech difficulty
- T: Time to call 911

Unfortunately, aspirin and CPR are not recommended here. While most strokes are caused by clots, some strokes may be caused instead by a hemorrhage in the brain. In such instances, aspirin would only worsen the problem. The best treatment for stroke is an immediate ambulance to the hospital.

The third type of cardiac event, while less commonly known than heart attacks or strokes, can be far more deadly: a pulmonary embolism (PE). In this case, a clot in the legs, usually a result of deep vein thrombosis, breaks off and travels to the pulmonary arteries, which deliver blood to the lungs. The American Lung Association estimates that about 900,000 people have a pulmonary embolism every year. As the clot grows, however, it may result in an acute or massive PE, causing the collapse of the right ventricle in the heart. Overall, these acute pulmonary embolisms account for over 300,000 deaths in the United States alone, with about 10 to 30 percent dying within one month of diagnosis. The warning signs of PE and an impending collapse include coughing, shortness of breath, pain in your back, and blue lips or nails. The best chance of survival is to catch it before it becomes an acute or massive PE, and work with a doctor as soon as possible to reverse the clot.

All three of these killers are the result of basically the same biological process of plaque building up throughout the blood vessels, rupturing, and clots forming; the only difference is where they occur in the bloodstream. And while overall efforts to respond quickly and with better lifesaving interventions are improving, in an ideal world, they wouldn't be necessary at all. Though the numbers may seem dire, the very good news here is that, with the lifesaving medications, advancements in medical imaging, genetic testing, and knowledge of healthy lifestyle that have developed in recent decades, we have the chance to completely eradicate atherosclerosis and the resulting thrombosis, therefore eliminating three of the leading causes of death in our country. And not only that; but as testing and imaging becomes more common, we have a greater chance of helping to prevent two of the other leading causes of death: cancer and Alzheimer's, as we'll discuss more in Chapter 10.

Detecting and Reversing Atherosclerosis

After working with Dr. Eliot in Phoenix, I became even more determined in my efforts to attract more attention to the idea of detecting and reversing atherosclerosis in its early stages. In addition, I now possessed unique expertise in the evaluation and treatment of psychological stress as a cardiovascular risk factor.

But I hadn't forgotten my dream to work for Dr. Kenneth Cooper.

A year after joining Eliot in Phoenix, Cooper called and said they were finally ready for me, offering a $50,000 raise. Eliot countered, and so I stayed on with him at their new location in Denver. But Cooper was always somewhere in the back of my mind.

A few years later, I was finally on my own, running my own clinic and traveling all over the world lecturing about heart disease prevention and all of the innovative testing that we had pioneered. I flew to Vienna, Rio, Manila, Frankfurt, London, and Hong Kong, talking about everything from stress and hot reactors to cholesterol and treadmills, and the new medications that were just arriving on the market that were expanding our ability to treat and reverse heart disease before any symptoms emerged.

Once, when I was in Argentina speaking about cardiac rehabilitation, I learned an important lesson about confidence. The doctor who had performed the first-ever bypass surgery, Rene Favaloro, MD, arrived at our conference like Juan Perón, wearing the fanciest suit you've ever seen, to the warm and admiring welcome of everyone there. It was almost like Jesus entering, with a crowd waving palm branches before him. There had never been any proof that bypass surgery prolongs life, except in rare circumstances like severe blockages of the left main coronary artery, so while the actual procedure of coronary artery bypass graft (CABG) surgery is undeniably a tremendously complex and skillful feat, I sat in the audience firm in my belief that our goal as a medical community should be to do as few bypass surgeries as possible. Today, we are on the verge of eliminating the need for CABG with the incredible new medicines that have arrived on the market.

But Favaloro's confidence inspired me in another way. He taught me that you need to be confident and hold your head high if you believe in what you are doing. The only reason bypass surgery became the standard treatment that it is today is because men like him would not be dissuaded. I realized that if I wanted my message to take hold, I would have to be similarly confident. I also recognized the need for humility: if I became wedded to certain ideas or treatments just because I was good at them or because they were lucrative for me, I would lose my mission entirely.

One thing I've learned is that when you're on the cutting edge of modern medicine, whether it's Cooper with treadmills, Eliot with mental stress testing, or me through most of my career, you will attract critics. After all, as physicians, we are dealing with people's lives. There are no higher stakes. Doctors should be challenged and criticized; otherwise, we risk arrogance, or we become blinded by our expertise. I welcome debate and embrace challenges. It makes me double check what I am doing, and then spurs me to keep going. Whenever I am discouraged and feel that we aren't making enough progress, or when I face intense criticism, I remember the American labour union advocate, Nicholas Klein's famous quote: "First they ignore you. Then they ridicule you. And then they attack you and want to burn you. And then they build monuments to you."

I loved what we were building in those days. I had a great staff, and we were working like dogs, but I was often gone and always paid myself last. It was a big sacrifice for Colette, who needed stability for our growing family, now three sons and a daughter. But I never forgot my dream to work with Cooper and become his protégé. In fact, I already viewed myself as something of a son figure to him.

Then one day, he called again. They wanted me to come to Dallas. I remember a tearful night with my wife, when I told her, "I just cannot do this by myself." Nobody was running the show. I am great at some things and terrible at others, so I was spending half my day just trying to work through administrative problems with my staff, and continuing to give too much money away. It wasn't working. Finally, it was time to join Cooper in Dallas.

We arrived on August 1, 1992, a day that is usually beautiful in Denver. But when we arrived in Dallas, it was horrible: 115 degrees in the shade. You couldn't lie down in the grass without bugs and snakes getting on you. As soon as we arrived, Colette said, "We're not staying here." She kept our dentists, doctors, neighborhood, and affiliations in Denver, and after a year of working with Cooper, my family moved back, while I began to commute back and forth, having given Cooper a two-year commitment.

If my home life was a bit chaotic in those years, my medical work reached new heights. I went to work full bore with Cooper and loved every minute of it. I had a unique skill set and expertise on his team, and soon became one of Cooper's right-hand men. When I arrived, Cooper's team of about eight doctors was still just doing treadmill tests, but I soon added a stress echocardiogram to the treadmill regimen, which enhanced detection of more serious blockages beyond just the EKG interpretation alone. Treadmills only discover the very worst plaque, detecting only a 70 percent or greater blockage in one or more of the largest coronary arteries. (My late father-in-law passed a treadmill test in the year he died!) If you have an abnormal treadmill, it's a sign that something is very wrong indeed, and you might have easily died for the last decade as the plaque, causing the blockage was growing before it matured and blocked the artery. But most cardiac deaths and myocardial infarctions occur as the plaque is growing. Treadmill testing or waiting for chest pain don't allow us to find it early enough. We've got to find plaque earlier. Three out of four heart attacks occur in coronary blood vessels that wouldn't have their smaller blockage detected on either electrocardiographic (EKG) or echocardiographic (stress echo) or nuclear (thalium) treadmills because they are designed not to detect plaque, but to detect ischemia (low blood flow). It's as though we designed the entire system to detect a fever (low blood flow, chest pain) without ever looking past it at the infection that caused the fever. To get at the root cause of heart disease, we've got to image the heart and arteries in order to find plaque in earlier stages, catching it before a cardiac event occurs without warning in a patient with no previous symptoms.

The idea of calcium scoring had also been introduced by Douglas Boyd in 1986, and then perfected by Arthur Agatston, creator of the South Beach Diet. I told Cooper about the possibility of calcium scoring to help us identify atherosclerosis earlier in patients, even those with normal treadmills. He was convinced; he bought a CT scanner and hired a radiologist to read it right away. We became one of the first centers to offer calcium scoring as a means of detecting and preventing heart disease. Our reputation grew, and we saw patients like Ross Perot, Tom Landry, and Stephen Covey—senators, athletes, and authors.

Cooper loved all the ideas I was bringing to the table, and tried to convince me and my family to stay, but our hearts were already back in Colorado. To counter, I tried to convince Cooper to start a Denver branch of his clinic, but to his credit, he wanted everything to remain under one roof in Dallas, and after I left, they expanded their attention to more holistic prevention of not just heart disease, but also cancer, diabetes, and other common diseases. Plus, my own speaking career was continuing to attract new business and increase my reputation as a preventive cardiologist in my own right. I returned to Colorado and started building what would eventually become the Boone Heart Institute. I brought the lessons and ideas I had learned from Eliot, Cooper, and many others together to create a framework for detecting and reversing atherosclerosis, which I'll introduce you to in Chapter 9.

It might strike you as odd that, although this entire book is about preventing heart disease, until now we've spent relatively little time talking about the heart itself, the central organ in this process. Although atherosclerosis (and the resulting thrombosis) is the number-one deadliest type of heart disease, most of the deadly action happens in the heart's blood vessels. The cardiovascular system is composed of not only the arteries and veins that feed the heart muscle nutrients and oxygen, but also the muscle itself, whose four chambers tirelessly move blood throughout the body, beating nearly one million times per week. The blood being transported to and from the body passes through four chambers whose entrances are patrolled by four pressure valves that open and close on demand when the heart muscle receives the electrical signal to contract.

Like any other muscles, our hardworking and seemingly tireless heart is also prone to fatigue and wear, both to the muscle itself and to its unique valves. The next chapter takes a look at these mechanical marvels, and what we can do to protect and preserve them.

CHAPTER 4

Wear and Tear: the Heart Muscle and Valves

Cameron takes a deep breath as he opens the door to the gym and steps out onto the basketball court. Everything is familiar—the same hoops, the same lines and scuffs on the shiny court, the same bleachers where his parents have sat for countless games, the same echo of the rubber ball reverberating throughout the gym, and the same jersey and his favorite shoes. This is the place where he has trained for the past three seasons on the varsity team at his Division 1 university, leading his team to countless victories even as a rookie. And yet, today is different.

Last week, Coach had called him to say that an NBA scout would be here today, watching him to see if he has what it takes.

It's the chance of a lifetime.

That must be why he didn't sleep well last night, and why his heart feels strangely fluttery and his palms are sweaty. Usually, Cameron swaggers out onto the court. At six foot eight, he is on the short side for a center, which he makes up for in tenacity and power. That was why he received a full ride to one of the best D1 schools in the country. Last year, he was named the MVP, and this year he's worked harder than ever to prove to himself and others that he deserves the title. With help from the team's nutritionist, he's been following a strict diet to make sure he receives all the nutrients he needs to maintain his high energy. He works out constantly, even before and after practices and games, often waking up at 5 a.m. to run on the treadmill. There is no room for mistakes. He is tired, yes, but that is normal. *You have to push through the pain. All of the greatest players do*, he tells himself.

As they begin their drills, Cameron tries to shake off the nerves, eyeing the bleachers where the coaches stand with another man he doesn't know, whose arms are crossed with a clipboard tucked underneath. He can feel the man's eyes following him. This is the moment he's been waiting for.

So why is he struggling? He takes an easy shot from ten feet. The ball hits the rim and bounces off. Cameron's face grows warm. He should have had that. Frustrated, he catches the rebound and dodges a teammate, spinning deftly around him to get another clear shot.

Come on, Cameron. You got this.

He spots his opportunity. Pushing hard, he propels himself up toward the basket for a clean hook shot—his specialty. The ball swishes through the net with a satisfying *whoosh* and his teammates cheer. Cameron holds his fist up in victory as he lands, suddenly feeling dizzy. The game is moving on, the basketball already down the other end of the court, but Cameron staggers, trying to regain his footing. His eyes blur and he suddenly can't breathe. Faintly, he can hear Coach shouting his name as he collapses on the ground.

"Get the AED! Call the doctor! Cameron's down!"

The competitive energy of the court turns to chaos as his team realizes something is wrong.

Cameron hears the whistle blow just as his world goes dark.

Unbeknownst to Cameron's doctors, coaches, and teammates, his heart has been working overtime for years. Despite being a shorter center than other players in the NCAA and NBA, Cameron is still a large man. You don't see many 250-pound seventy- or eighty-year-olds walking around at the beach in retirement—and there's a reason for that, which we'll discuss later. Cameron's heart, like many large men's, works hard to pump blood throughout every inch of his powerful frame. Not only that, but after thousands of practice sessions and drills, pushing himself at elite levels, the walls of his heart have thickened just as any muscle would. On this particularly stressful day, Cameron's tireless heart begins to beat too quickly, the muscle quivering with a rapid pulse that sends Cameron spiraling into a sudden cardiac arrest. Hidden high blood pressure, unmeasured heart thickness, excess caffeine, adrenaline from stress, and genetic tendencies have created a hidden storm inside of Cameron's heart muscle.

In someone so young, so talented, and so strong, most people never see it coming.

To be fair, failures of the heart muscle, and its inner maze of valves and electrical cords, are rare. Congenital conditions and extreme wear are the most common reasons this miraculous organ might malfunction. All organs age, even in healthy individuals. The seemingly indefatigable heart will begin to falter with "normal" wear and tear over the course of decades, unless it is under a watchful eye. In the past two chapters, we discussed atherosclerosis and thrombosis, by far the most prevalent manifestations of heart disease. But these are still just two of the four types of heart disease that we are trying to prevent at the Boone Heart Institute. In this chapter, we'll learn more about the heart itself and the incredible resilience of our most important muscle.

Righting Our Past Misconceptions

It's no wonder that the heart has been a symbol of love and devotion throughout history. You'd have to look long and hard to find anything that works as hard as a heart. For the last two thousand years, scientists have fought to reveal the inner workings of the heart, both for its biological and symbolic importance.

In ancient Greece, the philosopher Aristotle, best known for his work in logic, also studied human anatomy and believed that the heart is a fundamentally spiritual organ. In studying chicken embryos, Aristotle found that the heart was the first organ to develop, and because it is centrally located in the body, he assumed it must therefore be the source of a person's intelligence and emotions. Aristotle's writings on this idea prevailed for the next five hundred years, and we still often refer to "matters of the heart" to note its metaphorical role in creating our emotions.

In the second century B.C., however, a doctor named Galen challenged Aristotle's notions. Galen attended to the gladiators of Pergamon, a city in what is now Turkey. Galen had a curious, if morbid, mind. Working with dying, often mutilated men, allowed him the opportunity to witness the inner workings of the body without having to dissect human corpses, a practice which had been outlawed by the Romans centuries before. What Galen learned about the body from them, as well as, unfortunately, from his experimental dissections and vivisections of animals, allowed him to successfully treat many people in ancient Rome, including his most famous patient, the Stoic emperor, Marcus Aurelius.

Galen was fascinated with the heart itself, which he also believed was connected to the soul, although he didn't believe, as Aristotle insisted, that it's our source of intelligence. Galen disproved Aristotle's theory when he witnessed a gladiator who was struck through the heart and remained conscious until he died. Galen reasoned that if the man could remain conscious despite the mortal wound, then the heart couldn't possibly be where our thoughts originated. What, then, was it for? Galen measured and dissected animal hearts, carefully assessing them to discover their secrets. In his book, *On the Usefulness of the Parts of the Body*, he described the heart as "a hard flesh, not easily injured. In hardness, tension, general strength, and resistance to injury, the fibers of the heart far surpass all others, for no other instrument performs such continuous, hard work as the heart." He was certainly right about that.

Galen was only partly right, however, about how blood flows throughout the circulatory system. Through his experiments in animals (and perhaps live observations of wounded gladiators), he was able to watch the blood's movement to the vital organs, particularly the liver, the lungs, and the heart. But understanding their functions was still something of a mystery. He theorized that blood is produced by the liver after we digest food; that the liver sends blood to the heart via the veins, and then distributes it throughout the arteries; and that the lungs are responsible for "cleaning" blood (hence the more vibrant color of blood after it leaves the lungs) before it is sent to the rest of the organs. According to Galen, blood was then absorbed in the tissues throughout the body, hence the need for the liver to constantly produce more. This also explained why we get hungry so often.

Just as Anichkov was unable to go far enough in his understanding of cholesterol to help us fully understand its role, Galen was close to the answer, but had misunderstood some pieces of the process, as often happens in medical research. Unfortunately, a more accurate understanding of the heart and cardiovascular system wouldn't be developed until more than fifteen hundred years later. Galen's observations and theories of biology dominated the field of medicine for the next millennia.

While most physicians of the seventeenth century accepted at face value Aristotle and Galen's theories about the heart and circulatory system, William Harvey had his doubts. As the personal physician to King James I, and later, King Charles I of England and Scotland, Harvey was influential and well-resourced. King Charles I was also a supporter of his work, providing him with access to the royal deer for conducting his own observations and experiments. By this time, research on human cadavers was no longer taboo, and Harvey set about dispelling many of the myths that had proliferated throughout the world for so long. As frustrated as Harvey was by Galen's ideas, I can't help but think Galen would have admired his pioneering spirit.

Harvey systematically knocked down Galen's theories, one after another. He proved that within the heart, blood does not pass through "pores" from chamber to chamber, but instead is moved from chamber to chamber in one direction by valves, much like doors that only open one way. He observed how the blood circulates through veins and arteries, and measured the volume of our blood, concluding that there was simply too much for the liver to be constantly producing blood from our food; instead, he reasoned, blood must be continuously recycled throughout the body. He noted the synchronized timing of our pulse and the beating of the heart—a fact we now take for granted—and was the first to insist that it is the heart itself that pumps blood throughout the body.

Although many physicians of the time were initially resistant to Harvey's ideas, his influence throughout the British empire was widespread. By the time of his death, his theories were widely accepted as truth, overturning fifteen hundred years of misunderstandings in the course of one human lifetime. I often think that if Harvey was able to do so much in so little time, how quickly might we continue to overturn the old ways of thinking about the heart and heart disease?

The Hardest-Working Organ

Progress always begins with new knowledge. Thankfully, with the aid of modern technology, we've been able to unravel the mysteries of the heart at the cellular level. Let's start with a more thorough understanding of this most vital of organs: the heart itself.

The heart is composed of two visible tissues: the muscle, which is divided into four chambers, and the valves: the gates that safeguard the entry to these four chambers of life, as shown in Figure 2.1. Each chamber of the heart plays a different role in the process of circulating blood and preparing it to be sent throughout the body. The upper two chambers, the atria, are responsible for taking blood in, while the lower two chambers, the ventricles, take the role of pumping blood out. The difference between the left and right sides is about where the blood goes.

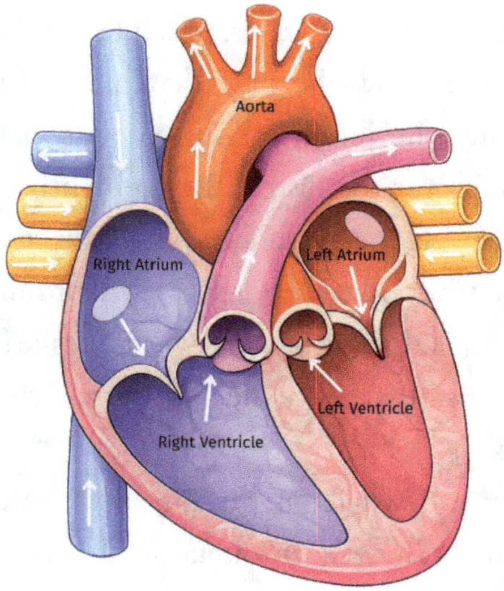

Figure 2.1

If you're following the path of blood through the heart, it arrives via two large veins (one from the upper body, the *superior vena cava*, and one from the lower body, the *inferior vena cava*) that deposit the blood into the right atrium after it has delivered oxygen and nutrients to the rest of the body. The blood, which is thick and dark after having been depleted of oxygen, then passes through the doorway of the tricuspid valve (a valve with three flaps that open to allow blood through) below to the right ventricle. With each pump, the right ventricle then sends the blood up through the pulmonic valve to the lungs to receive a fresh supply of oxygen. (We can understand where Galen got the idea that the role of the heart is to support the lungs!) The now-reoxygenated blood is a vibrant red and returns through the pulmonary veins to the heart, where it is received by the left atrium, which passes it through the mitral valve to the left ventricle. The left ventricle then sends blood out to the body via the aortic valve to the aorta, the largest artery. The entire cardiovascular system is a closed circuit, with the arteries distributing oxygenated blood throughout the body, and the veins transporting deoxygenated blood back to the heart, where the cycle begins again.

The filling of the left ventricle is a particularly good indicator of the overall health of the heart. As you can see in Figure 2.1, the ventricles of the heart are lopsided, with the left ventricle appearing much bigger than the right one. This is because the right ventricle's only job is to send blood to the lungs, just around the corner about an inch away. The left ventricle, however, is responsible for sending blood throughout all sixty thousand miles of blood vessels, from the central organs all the way up to the brain and out to the tiniest vessels and capillaries in our toes.

If the heart is the hardest-working organ, then the left ventricle is undoubtedly its hardest-working chamber. Therefore, the muscular walls of the left ventricle are much thicker than the walls of the other three chambers. No doubt, when Galen admired the strength and resilience of the heart muscle, he was looking particularly at the left ventricle. Like all muscles, the left ventricle gets thicker when it works hard. This is called hypertrophy. The thickness can be physiologic (normal and caused by healthy exercise) or pathologic (caused by disease or excess work straining the heart). Physiologic hypertrophy is associated with a big, strong, healthy heart which, despite its size, remains supple and flexible. Pathologic hypertrophy can also cause the heart to look big and strong when the muscle is actually stiff and inflexible, often due to high blood pressure, stress, and overwork, overlaid onto a matrix of genetic susceptibility. A simple echocardiographic test separates good (physiologic) from path (pathologic) hypertrophy. The test is called an E/A ratio and it measures heart stiffness by assessing how easily and smoothly the heart muscle rapidly empties and fills. An E/A above 1.10 is good. A ratio below 1.10 is cause for concern, and indicates a need for further evaluation and treatment. We use this simple echocardiographic measure to sort out the cardiac health and safety of professional athletes. But every individual needs this, too. We want hearts that look like Michael Jordan's, not Arnold Schwarzenegger's.

With each beat of the heart, the muscles contract and expand in a motion similar to wringing out or squeezing: an intricate dance of electrical fibers and cells working in coordination with each other. Ideally, the heart (particularly the left ventricle) is supposed to uniformly pump out 50 to 75 percent of the enclosed blood with each beat, a measurement known as the ejection fraction (EF percent). You might think that the higher the ejection fraction, the better, but if your heart were to squeeze out 90 percent, it would be a sign that the stiff left ventricle couldn't expand enough to accept blood from the left atrium—in other words, it would have a low E to A ratio. On the other hand, a low ejection fraction means that the heart is failing to pump blood at a high enough volume and velocity out to the body. At one point, when my father was in his final days, his ejection fraction was a mere 19 percent, barely squeezing out any blood at all. The pump was failing at age ninety-eight.

The valves also play a special role in maintaining this synchronized flow of blood throughout the organ. Their primary purpose is to make sure that the blood continues flowing in the right direction, from right atrium to right ventricle, a quick detour to the lungs, back to the left atrium and left ventricle, and then out to the arteries. You can think of the ventricles as the largest rooms in the house: they each have a door that allows blood to flow in, and a door that allows blood to flow out. As blood fills each ventricle from the atria, the "in" doors (the tricuspid and mitral valves) open, while the "out" doors (the pulmonic and aortic valves) remain closed to allow the ventricles to fill to capacity. With the next pulse, the "in" doors close and the "out" doors open as the ventricles pump the blood to its next destination. With this alternate opening and closing of the valves, they make the soft *lub-dub* sound that doctors listen for through their stethoscopes.

For most of us, these tireless pieces of hardware will continue to open and close, open and close, without issue throughout the majority of our years—unless, of course, a problem exists from the beginning.

Congenital Heart Defects

In 2014, Denver Broncos head coach John Fox collapsed while playing golf in Charlotte, North Carolina, the result of a stenosis, or narrowing, of his aortic valve. The aortic valves are also tricuspid, meaning they have three flaps that work together to pump blood in the right direction. However, John's aortic valve was bicuspid, with only two flaps, and the risk for malfunction was high. John had known about his congenital condition for decades, and yet, he reasoned, his heart had been operating just fine all this time, despite the risks. I was the preventive cardiology consultant for the Broncos, and together with his team of doctors had tried to persuade him to have surgery during the prior offseason to address the issue. He refused. Finally, with his wife's urging, John agreed to have surgery in the *next* offseason. I can understand Coach Fox's decision to delay; most of us in his position would probably do the same. After all, he felt fine, just as he had for decades.

Unfortunately, his heart couldn't wait. He was golfing with friends during his team's bye week when he collapsed suddenly. The Broncos were in the middle of a Super Bowl season and their head coach needed immediate heart surgery to replace the faulty valve. The surgery was a success, but the delay came at a cost. Less than three months later, when John was not yet fully recovered, the Broncos played the Superbowl game in New York City and lost to the Seahawks. Would that game have ended differently if John had had the surgery eight or nine months before and had more time to recover?

Congenital heart issues are often hard to diagnose, and much like atherosclerosis, they seem to strike without warning. Returning to the analogy of the house, sometimes you can go a long time with a door that doesn't quite latch properly, or a foundation that's sinking a bit. But eventually, if left unaddressed long enough, these problems can become a threat to the entire structure.

If you follow college sports long enough, you'll hear of another lethal congenital condition called hypertrophic cardiomyopathy. With this condition, the walls of the left ventricle are unnaturally thickened. The heart's electrical fibers can become tangled within this web of abnormal cardiac muscle fibers. This is surprisingly common; hypertrophic cardiomyopathy affects about one in five hundred people and is the most prevalent congenital heart disease. As Cameron experienced, when someone with hypertrophic cardiomyopathy is exposed to extreme stress through exercise, the heartbeat speeds up to almost a quivering pace in what's called ventricular fibrillation—technically an electrical issue, which we'll discuss more in Chapter 5—causing chest pain, dizziness, shortness of breath, and a rapid heart rate. If not treated immediately, the victim will fall into "sudden" cardiac arrest (SCA), in which the heart stops beating altogether. Over 210,000 Americans die each year from SCA, and hypertrophic cardiomyopathy is a leading cause in the United States.

Hypertrophic cardiomyopathy has been responsible for cutting short the careers of many promising young athletes, including that of college basketball legend Eric "Hank" Gathers. At just twenty-three years old, Gathers ranked at the top of all players in Division 1 schools. Playing forward for Loyola Marymount, he seemed unstoppable. However, in December 1989, Gathers suddenly collapsed during a game against UC Santa Barbara. He was diagnosed with an irregular heartbeat, but soon returned to the court and seemed to be fine. Unfortunately, just a few months later in March 1990, he collapsed again during a West Coast conference semifinals game. He died at the hospital two hours later.

Episodes like this are all too common, especially in stop-and-start sports like basketball (which has the highest incidence of fatality from hypertrophic cardiomyopathy), soccer, football, and track and field. These sports rely on short bursts of remarkable energy, putting an immense amount of strain on the heart muscle. Not only that, but the athletes who play these sports are typically tall and large in stature, adding an extra burden on the heart, and they train at increasingly elite levels in competitive high school and college sports. Preventing the deaths of young athletes is the primary reason we founded the EveryHeart Foundation. We are working hard to ensure that testing is affordable and available for every student athlete in the nation. As further evidence of just how necessary this testing is, in July 2023, LeBron James Jr., son of the LA Lakers superstar, collapsed from an undisclosed cardiac condition while preparing for the start of his college basketball career at the University of Southern California. Thankfully, he recovered from the incident, but if we had been able to test him, we might have been able to prevent it from happening in the first place.

Occasionally, when we test student athletes, we find a small hole in an athlete's heart, called a *patent foramen ovale* (PFO). All of us are born with a tiny hole through the atrial septum. For most people, the heart continues to form and this hole closes as we grow older. For those with PFO, however, a hole remains into adulthood. According to the American Heart Association, about one in four people, or 25 percent of us, have this condition. Depending on the size of the hole and the person's lifestyle, it may never cause any problems, or it might eventually cause a stroke or migraine headaches. Generally, it's not diagnosed unless something else goes wrong first.

Other structural issues can appear if the heart doesn't develop fully in utero. In addition to PFO, a hole might develop if the atria or the ventricles neglect to form completely, causing an atrial septal defect (ASD) or a ventricular septal defect (VSD). And both of them can lead to trouble.

My daughter, Kiki, was the captain of the University of Denver lacrosse team, leading her team to the NCAA tournament for the first time in 2006. The program continued to ascend and eventually the Pioneers reached the final four in 2023. One of her teammates was from the East Coast and had always believed she had asthma. For years, she took various asthma medications, and when she moved to the high altitudes of Denver, her breathing grew worse. Fortunately, my daughter brought her to the Boone Heart Institute and we found that she had an ASD. All her life, she had been treated for the wrong problem. She received surgery to repair the hole, transferred to a college in Philadelphia, and continued to play lacrosse, without the need for her asthma medications.

Other possible structural defects in the heart include problems with the valves. If you've heard of someone being diagnosed with a heart "murmur," it means that instead of the typical *lub-dub* sound that doctors expect to hear when checking someone's heart, the flow of blood instead creates a soft whooshing sound as it passes through the valves. This disturbance in the blood flow can be caused by stenosis, which is the thickening and narrowing of the valve. Stenosis often occurs with a bicuspid rather than tricuspid aortic valve, like the one that John Fox was born with. Or it may be due to an excessive lesion or buildup of plaque from atherosclerosis. The other problem is regurgitation, or the leaking of blood both ways through the valve. In other words, blood flows back through the valve into the chamber that it's supposed to be leaving. With either issue, the ventricles or the atria may be receiving either too much or too little blood as the valves perform inadequately, creating extra work as the blood moves to its next destination. These valvular issues may lead to a breakdown of cardiac function, and necessitate a surgical valve repair or replacement.

The Dangers of Extreme Exercise

Legend has it that in ancient Greece, a courier named Pheidippides ran twenty-six miles from the battle of Marathon to Athens to announce the Greeks' victory against the Persians, and then promptly died. There's no record of this event occurring; this legend comes not from Greek mythology, but rather from a poem by the nineteenth-century English poet, Robert Browning. In the poem, after the Persians are destroyed, Pheidippides "flung down his shield / ran like fire once more... / Till in he broke: 'Rejoice, we conquer!' Like wine thro' clay / Joy in his **blood bursting his heart**, he died—the bliss!" (emphasis added)

It's a fitting and memorable story cautioning against excessive exercise, and surprisingly accurate in describing the intense strain on Pheidippides's heart, although, if Pheidippides could talk, I don't think he would describe it as "bliss." In terms of biology, it's likely that Browning was describing an aortic rupture, or a tear in the lining of the aorta, usually caused by extremely high blood pressure after bouts of intense physical activity, and nearly always fatal. Or maybe the cause was an electrical storm in the heart, causing sudden cardiac death due to adrenaline-driven explosion of cardiac muscle cells swimming in an oxidized, dehydrated pool of thickened blood.

It may seem odd that excessive exercise should be such a prominent risk factor for heart disease. After all, isn't exercise, along with good diet and nutrition, one of the best ways to defend *against* heart disease? Isn't stasis, the lack of movement, one of the primary risk factors for thrombosis?

The benefits of exercise are indeed well documented. Regular, moderate exercise can strengthen the heart, resulting in improved blood flow and higher oxygen levels throughout the body. It lowers blood pressure and increases the natural number of HDL cholesterol in bloodstream. Exercise also lowers the risk of type 2 diabetes, another risk factor for atherosclerosis; increases our mobility; and decreases the risk of clots, colds, and inflammation. Undoubtedly, most of us need *more* exercise in our lives, especially as we get older.

But the advantages of exercise have a limit. In the 1990s, Dr. David C. Nieman at Appalachian State University found an interesting relationship between the risk of upper respiratory tract infections (URTIs, aka the "common cold") and exercise. He charted this relationship in an illustration now famously known as the J-shaped curve, shown in Figure 4.1.

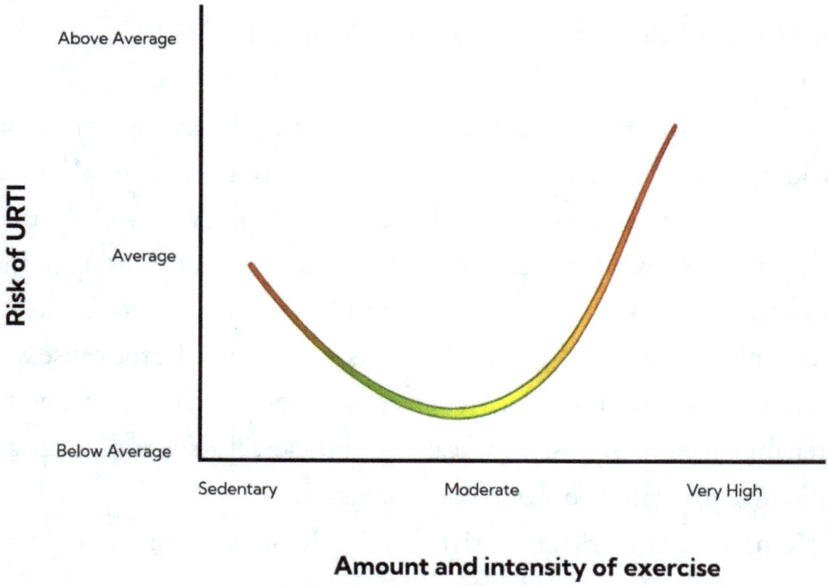

Figure 4.1. Source: Nieman, 1994.

Nieman wrote of the J-shaped curve, "This model suggests that while the risk of URTI may decrease below that of a sedentary individual when one engages in moderate exercise training, risk may rise above average during periods of excessive amounts of high-intensity exercise." In other words, we get better when we engage in regular, moderate exercise, but far worse when we engage in too much. Since Nieman's original paper, the J-shaped curve has been shown to apply not only to the risk of URTI, but also to other health conditions, including the health of the heart.

Dr. Ralph Paffenbarger, Jr., documented the same phenomenon in Harvard alumni, finding that those performing more exercise exhibited better health outcomes, but only up to a certain limit. Those burning more than two thousand extra calories per week with aerobic exercise were actually *less* healthy than their more moderate counterparts.

Remember Jim Fixx, who popularized the jogging movement in the 1980s and died of a heart attack during a warm afternoon run? Not only was he genetically predisposed to atherosclerosis; he was also obsessed with the sport. He was what Dr. Alayne Yates called an "obligate runner"—someone for whom exercise is almost a compulsion, to the point that it has a destructive impact on their health. Our young basketball star Cameron may fall into the same category. Fixx was known for touting the benefits of running and other exercise, and is responsible for a positive shift in the public consciousness towards increased exercise. Many runners who took up the sport did indeed find, as Nieman's work shows, an initial decrease in their instances of colds and other common ailments. There's no doubt that most of the people who began jogging due to Jim Fixx's encouragement, experienced positive health outcomes. But for Fixx himself and many other athletes who push themselves to extremes, the damage done outweighs the possible benefits. I like to emphasize to my patients that health is different from fitness. Severe exercise may be good for your fitness, but don't be deceived into thinking that it is good—or necessary—for your long-term health. Marathon-running doctors like Dr. Paul Thompson of Rhode Island, Dr. Vogel of Colorado, and Dr. Peter McCullough of Dallas have helped quantitate the dangers of "excessive endurance exercise," often known as E3 or triple-E.

Denver, Colorado, where I live, is full of ambitious and dedicated athletes—serious runners, rock climbers, and hikers, who are health-conscious and imagine that they are in excellent condition. I myself have run a few marathons. But extreme, competitive exercise may place too much of a demand on the heart muscle, leading to problems like atrial fibrillation and cardiomyopathy. I often worry about the fittest, strongest athletes because they think they're fine, and they're often not. This is another reason we test young athletes for existing conditions; because these particularly vulnerable children are often pushed to the absolute maximum that their still-developing hearts can handle. But the weekend athlete or club fitness champion needs a quick look as well, to assess how their anatomy, biochemistry, and genetics interact with their fitness habits and goals.

Size Matters

As we mentioned earlier, the physical size of the heart in proportion to the size of the body is a factor in extreme exercise. This is a risk for particularly large adults, including many of the supersized athletes I work with. We find that big, strong athletes often dominate until about age fifty. After that point, their bodies tend to wear down quickly. These athletes' hearts must pump blood throughout a larger area and are regularly called upon to perform in extreme circumstances. NBA player Kareem Abdul-Jabbar, at seven foot two, is a prime example. After an impressive career in the NBA, Kareem was diagnosed with atrial fibrillation (AFib), a condition in which the atria seem to quiver or vibrate rather than pump cohesively, similar to the ventricular fibrillation discussed earlier, though the effects are often intermittent and less acute. Large, well-trained hearts often develop atrial fibrillation later in life. Nearly every four-minute miler in American history has suffered from AFib because the incredible power required to run so fast causes their hearts to become huge, performing well but stretching their electrical capacity. Unless you're the Grinch, having your heart grow three sizes in a short period of time is a major cause of concern.

In the short term, AFib causes dizziness and trouble breathing, and it can be easy to ignore if you're used to high-intensity workouts. If left untreated, however, it can cause major problems such as the formation of a clot inside one of the atria. If that clot were to break loose, it could cause a stroke. Since his diagnosis, Abdul-Jabbar has become a spokesperson for the heart drug Eliquis, which is used to prevent blood clots and strokes and has been lifesaving for many elite athletes and others aging with atrial fibrillation. Large individuals and endurance athletes are at earlier risk, but all of our aging hearts will experience a deterioration of the electrical wiring in the heart. High blood pressure, obesity, and atherosclerosis will fray the electrical fibers, as you'll learn in the next chapter. Stress and alcohol can also trigger events of atrial fibrillation. Controlling these various risks will delay and prevent AFib.

The impact of the sheer size of the heart muscle, especially when under stress during extreme exercise, has also been well documented in other large mammals, including horses and whales. A 2021 study of just under five thousand racehorses found AFib to be present in nearly 5 percent of the horses. That may not sound like a lot, but is high when you consider that none of the horses had any underlying or preexisting structural conditions. In other words, if not for the rapid increase in their heart size and the constant stop-and-start stress on their hearts, these horses would be perfectly healthy. For small mammals like mice and voles, this is never a concern.

Old Age

Even in people without congenital conditions who rarely participate in extreme exercise and are of average height and weight, the muscle and valves of the heart can simply wear down over time. If atherosclerosis and thrombosis don't get you, heart failure as the result of normal wear and tear of our hardware is the next likely culprit.

There's an old myth that everyone's heart is allotted a set number of beats in its lifetime, and that once you've used them up, your time here on earth is done. This is untrue, but the average person's heart will beat an impressive total of more than 2.5 billion times in their lifetime. I mentioned in the Introduction that by the time my father passed, his heart had beat nearly 3.5 billion times. It's no surprise, then, that he had to have a valve replaced in his early nineties. Like the hinges on a door wearing down over time, sometimes the valves simply wear out.

Similarly, a patient of mine came in to see me complaining that he was short of breath. It was actually much worse than that. We found that one of the cords, or tendons, that holds one of his valves (the atrial valve) in place had ruptured, so that the valve was flapping in the bloodstream, much like a boat's sail flapping in the wind. The valve was repaired and two decades later, this patient is still enjoying an active lifestyle. Incidentally, the surgeon who repaired the valve was our neighbor who died while running a marathon in San Diego a few years later.

That surgeon's partner, Myles Guber, also a valve surgeon, is now a patient of mine and speaks every year at our conference: The Summit: Preventive Cardiology in the Heart of the Rockies. He jokes that if everyone was on the Boone Heart Protocol, he would have to shut down his practice.

Fortunately, we have a test that can conclusively diagnose heart failure before victims suffer from this chronic condition. This new magic heart-failure predicting blood test is called NT-proBNP (N-terminal pro-B-type natriuretic peptide). NT-proBNP is a biomarker, a substance in the blood, that helps doctors assess heart function, particularly in cases of suspected heart failure. Elevated levels of NT-proBNP indicate that the heart is under stress, often due to conditions like heart failure or other cardiovascular issues.

A healthy score for the NT-proBNP test would be under 125 ng/L. Years ago, I was working on patient charts during a study break at a heart failure conference in La Jolla when the pioneer of NT-proBNP, Allan Maisell, MD, walked past the table where I was sitting. He noticed that my patient had a NT-proBNP score of 135. I told him we measured NT-ProBNP in every patient every year, and then treated the abnormal ones. He said, "No one will ever get heart failure in your practice." If we could find a way to maintain a low NT-ProBNP score throughout our lives, our hearts would never fail. Any number higher than that indicates that the heart is straining to pump. I've seen this number as high as thirty thousand in an ex-NFL player who needed a heart transplant. My father had a score of sixteen thousand a few years before his passing. Thankfully, the heart is resilient, and we were able to successfully strengthen the muscle with medication, lowering his score to four thousand—still high, but low enough to sustain him for a while longer.

Structural damage to the heart that causes increased NT-proBNP can be serious, but if caught early, there's much that can be done. In the most dire cases, someone might need major surgery to repair any damage done, but on the whole, it's reasonable to think that if we can keep our heart muscle strong, we might easily live to a hundred or more. And even then, the cause of death wouldn't be a leaking valve or a thickened left ventricle, but more likely outliving our own skeleton or other organs. There are other experts hard at work to find out how to extend the lives of those organs as well.

There is one more category of heart disease that we try to prevent at the Boone Heart Institute. Returning to the house analogy, when the structure is sound, the doors open properly, and the pipes are clear, we might be tempted to think all is well and good. But there's one other area to look out for: electricity. You never know when, seemingly without warning, the lights will begin to flicker—or go out altogether.

CHAPTER 5

When the Lights Go Out: Electrical Issues

Clara taps her toes nervously as she sits on a plastic school chair, waiting for her turn behind the makeshift privacy screens that have been erected in the gym. Her school has arranged for all the student athletes to be tested for heart defects and other issues. She was given an appointment time and told to come to the gym. She's already heard about another boy on the varsity track team who apparently has a small hole in his heart that will require surgery. But what are the chances of something like that happening? Everyone else so far has been fine. They won't find anything on her. Right?

As the goalie of the girls' junior varsity soccer team, Clara is proud of herself. She loves soccer: the smell of the freshly mowed grass and the positive energy of all the girls on the team working together. She loves that she gets to work out with her best friends every day. Her parents, both immigrants from Vietnam, have instilled in her a strong work ethic, and she's determined to make it onto the varsity team next year.

Finally, a nurse holds open the screen and Clara's teammate comes out, smiling. "Come on in!" the nurse calls to Clara.

"Don't worry, it's painless," her teammate promises her with a grin as they pass.

Inside the booth is a tall machine with a panel of controls and a large screen on top. The nurse asks her for some details: her name, date of birth, height, weight. Then she asks what sport and position Clara plays—just small talk. Clara answers quickly.

"Do you ever have any fainting or dizziness on the field?" the nurse asks, smiling in her blue-green scrubs.

"Well, I did faint a couple of times," Clara admits. During last year's game against their biggest rivals in the district, no less. But that had been a particularly warm day, and they'd all been pushing hard to beat their competitors. Probably dehydration, her coach had said. Then it happened again this year, when, in the middle of a practice, Clara suddenly felt like she couldn't breathe and had to sit down on the field. The coach had pulled her out of practice that day and had her take a few days off to rest. Since then she'd been fine, though.

"And you play goalie?" the nurse asks, with a slight frown. Clara nods.

"Okay, let's check it out."

The nurse leaves so that Clara can undress from the waist up and slip into a modest hospital gown. When the nurse comes back in, she walks Clara through the tests as they go. First, the nurse takes her blood pressure, which is normal.

"Now we'll do it again, but this time I want you to count backwards from 777 by sevens."

"Count backwards by sevens?" Clara asks, bewildered. She *hates* math. The nurse laughs at the expression on her face and nods.

"Ok, 777, 770, um ... 763 ... 756? Or 757?" Clara continues for a while as the nurse takes her blood pressure again.

"Much higher this time," the nurse tells her. "That happens sometimes."

Next, using the ultrasound machine, a female technician uses a small, gel-covered probe to inspect her heart and the arteries in her neck. Clara is awed, watching the screen's black-and-white image of her heart pulsing with each beat. The nurse looks intently at the image.

"Everything here looks normal, but our doctors will confirm that," she says. Clara can sense that she's not quite sure. It's almost like the tech is looking for something. Clara doesn't dare ask if anything's wrong. Her heart starts pounding harder.

The last test is an electrocardiogram (EKG). The nurse attaches little electrodes to Clara's chest, wrists, and ankles, and they watch the screen together. The beats of her heart are amplified by the machine and a small, scratchy line of sharp mountains and flat valleys appears. For several seconds, the mountains and the valleys follow each other in a steady pattern, until suddenly the pattern changes—the line suddenly fluttering with her corresponding anxiety.

"What does that mean?" Clara dares to ask.

The nurse takes a deep breath and turns to her.

"We're going to need to order some further tests, I think. This might explain your fainting. It's complicated, but might be simple for our doctors to fix."

Clara is like one in a hundred children we test through the EveryHeart Foundation who has a suspicious arrythmia, or irregular heartbeat. Even if a patient appears completely

asymptomatic, an arrhythmia could be a sign of a more sinister problem lurking beneath. Much like a sudden clot or embolism, the first symptom can be death. And for those like Clara who already exhibit some symptoms, like fainting, dizziness, and shortness of breath, and especially those who play a sport like wrestling, football, baseball, or soccer, where the chances of colliding with a stray ball or body are high, the sooner the underlying electrical problems can be deciphered, the better. Thankfully, for Clara and many other young children, with early testing we can catch these problems before they become a lethal threat. In Clara's case, our team identified a condition called Wolff-Parkinson-White Syndrome (WPW). We then sent her to a specialized cardiologist called an electrophysiologist, who treated her condition using tiny catheters to either freeze or laser the abnormal electrical pathway that was discovered on the EKG to return the heart to a normal rhythm, even during exercise. As you'll see, the issues may be genetic, or they may generate from a structural problem in the "wiring." Or there may be no identifiable cause at all.

The beating of the heart is one of the most fundamental biological processes that our bodies undertake every day, without us ever having to think about it. The average person's heart beats about 2.5 billion times in their lives, and yet, for a function so vital (literally, essential to life), we understand little about this fourth category of possible heart disease, compared to the first three areas. Electricity is the final frontier of cardiology. As preventive cardiologists, we can clean out blood vessels, identify clots and embolisms before they cause a stroke or heart attack, improve the muscle's flexibility and thickness, and replace the valves of the heart, but we're still learning how to keep the lights working in the house. The system can always short-circuit no matter how perfectly everything else is going. While atherosclerosis, thrombosis, and problems with the muscles and valves can be relatively easy to address, electrical problems can be harder to repair. Thankfully, if you do keep the rest of the cardiovascular system healthy, the electricity usually continues to operate normally as well.

But in order to anticipate and prevent the ways that the system might malfunction, we need to understand how it works in the first place.

The Jellyfish and the Tortoise

Only in the last 150 years have we begun to understand the process of how the heart generates electricity. It's no surprise that Aristotle, Galen, and many other past physicians believed that the heart must have some kind of spiritual significance; how else could it generate its constant rhythmic beats without a source of divine inspiration? God as watchmaker, winding us up and setting our ticking hearts into motion. Our friend, William Harvey, who debunked many of Galen's theories, was the first to propose that there must be a *biological* rather than theological reason for our beating hearts. In his 1651 book *Exercitationes de Generatione Animalium* (Latin for *Exercises on the Generation of Animals*), he wrote that "the pulse has its origin in the blood ... the cardiac auricle from which the pulsation starts, is excited by the blood."

By the nineteenth century, scientists knew that the heart's beat was not actually generated in the blood. A popular theory was that it must be triggered through the nervous system and ultimately generate in the brain, although that idea had been challenged in the late eighteenth century when the Swiss Anatomist, Albrecht von Haller, removed the hearts from animals and discovered that they continued to beat even after having been severed from the surrounding nerves. Haller concluded that the heartbeat must somehow be generated within the muscle itself, an idea called the myogenic theory. However, this too was challenged, when French physician, Cesar Legallois, found that a crushed spinal cord would cause a heart to stop beating—a clear indicator, or so it seemed, of the influence of the nerves on the electrical pulses of the heart, a hypothesis called the neurogenic theory.

In 1839, Czech doctor, Jan Evangelista Purkinje, identified thick fibers that surround the heart's two ventricles, which seem to carry the heartbeat, supporting the myogenic theory. These fibers were so thick and comprised of a substance different from other skeletal muscles that he believed that they must be made of cartilage, though later he admitted they were muscular. That same year, Polish neurologist, Robert Remak, discovered the presence of ganglion cells, or clusters of synapses that serve as the intersections between nerves at the top of the atria, from which the heartbeat appeared to start before travelling along the fibers that Purkinje had found, supporting the neurogenic theory. The debate between the neurogenic versus myogenic heartbeat continued.

Finally, in 1860, British doctor, Michael Foster, was determined to get to the bottom of the mystery. If scientists believed the ganglion cells must be the source of the heartbeat, there was one way to find out for sure: remove the ganglion cells and see what happened, which is exactly what he did. Using frog hearts, Foster cut away sections of the ventricles and atria, finding that the apical, or lower, section of the heart would not continue to beat, whereas the basal, upper section would. However, when he applied an electrical current to the apical section, it began beating again. If the heart was a normal muscle like any other, then this was an odd reaction; all the other muscles in the body would seize up completely when an electrical current was applied. Foster realized that the tissue of the heart itself was different—it seemed oddly predisposed to beat rhythmically, although he acknowledged the role the ganglion played as well. Perhaps *both* theories were correct?

The answer came from an unlikely source.

One of Foster's students was the biologist George Romanes. While you and I might spend our summer vacations lounging on the beach and drinking Mai Tais, Romanes spent his holidays on the Scottish coast studying jellyfish. Ever the biologist, Romanes was entranced by the beauty of these serene creatures, which he called the "Medusae," and took great pleasure in spending his summers observing them and conducting experiments. He noted that jellyfish swim via contractions of their upper dome-like muscle, which Romanes called the "bell." The bell has a rim of muscular tissue, which seems to control the contractions. Romanes found that when the rim was removed from the rest of the bell, the bell could no longer contract to allow the jellyfish to swim, although the rim itself continued to contract. And yet, that muscle did not appear to have any ganglion cells or nerve fibers that orchestrated the timing of the contractions. Romanes wrote that the rim was a unique blend of nerve and muscle, apparently beating of its own accord.

Another of Foster's students, William Gaskell, heard about Romanes's experiments and wondered if the same might be true of heart tissue. He decided to replicate the experiments, this time on the heart of a tortoise, rather than a frog like his tutor Foster had. Switching to tortoise hearts made the critical difference. Frog hearts only have three chambers: two atria and one combined ventricle that pumps blood to both the lungs and the rest of the body. Tortoise hearts more closely resemble human hearts, with a septum dividing the lower chambers into two ventricles.

Through a series of experiments, Gaskell showed that the contractions of the tortoise's heart flowed in a wave from the bundle of ganglion cells, called the sinus, at the top of the atria across the septum, and throughout the fibers around the ventricles discovered by Purkinje—much like a jellyfish's bell contracting as it swims. And yet, also like the peculiar rim of the jellyfish, Gaskell found that the ventricles will continue to contract even when separated by a "block" from the sinus. The debate was over. In 1883, Gaskell published his conclusions: that the heartbeat is a contracting wave of electrical pulses that emanates from *within* the muscle itself, and not from a neurological stimulus.

No doubt, equally influenced by his colleague Gaskell's findings, Romanes described the swimming of the jellyfish in these terms fifteen years later, in 1898:

"At each contraction of this muscular sheet the gelatinous walls of the bell are drawn together; the capacity of the bell being thus diminished, water is ejected from the open mouth of the bell backwards, and the consequent reaction propels the animal forwards. In these swimming movements, systole and diastole follow one another with as perfect a rhythm as they do in the beating of a heart."

Throughout the next decades, Gaskell's (and Romanes's) experiments were replicated with various other mammal hearts and his support for the myogenic theory was confirmed. In 1893, Swiss cardiologist and anatomist, Wilhelm His Jr., identified the bundle of fibers between the atria and the ventricles that conducts the electrical impulse running through the septum, now known as the Bundle of His. In 1906, Japanese physiologist Sunao Tawara confirmed the existence of the atrioventricular (AV) node, where the electrical pulse passes from the atrium to the ventricles. A year later, British anatomist, Arthur Keith, and British physiologist, Martin Flack, also identified the sinoatrial node (SA), from which the wave of contractions originates.

In the decades since then, as technology has improved and we've been able to examine the heart at the cellular level, we have gained a more holistic understanding of how this special blending of nerve and muscle conducts electricity. As the heart is responsible for pumping blood throughout the body, it relies on a steady beat generated by the sinoatrial (SA) node, located at the top of the atrium, to maintain its rhythm. This node sends out an electrical pulse, like a metronome, to signal the atria to contract, then travels down to the atrioventricular (AV) node to signal the ventricles to contract, resulting in the *lub-dub* sound doctors are listening for with their stethoscopes. The contraction is a movement called *systole*; the rest between beats, *diastole*—this is where we get the terms for systolic and diastolic blood pressure. When you're sleeping, the heart might beat at a reliable forty beats per minute; in waking activity it might range from sixty beats per minute while you're resting to upwards of a hundred eighty beats per minute while exercising. With each beat, the current of electricity is passed from cell to cell very quickly along this pathway, signaling the heart to contract—so quickly, in fact, that the cells of the muscle contract and release in near perfect unison.

In *The Song of the Cell*, Mukherjee summarizes the beating of the heart as a complex network of individual cells, but with the unique ability to contract and release in coordination with each other in time with the electrical pulse they receive from the SA and the AV nodes. "We might imagine the organ," he writes, "as if it behaves, almost, as a single-minded cell." We don't have to think about this pumping effort; it just happens. However, at any point in the transmission of electricity from cell to cell along this pathway, the signal can get disrupted.

Arrhythmias

A healthy heart at rest beats between sixty and a hundred times per minute, in response to the steady electrical pulse from the SA node, known as our internal "pacemaker." The disruption of these electrical pulses, however, will cause an irregular heartbeat called an arrhythmia. There are several types of arrhythmias that can happen, usually in response to a genetic or lifestyle trigger, or as a result of another disease. Remember, most electrical issues are secondary issues; they're a sign that somewhere, something is amiss.

Depending on where and how the disruption occurs, the atria or the ventricles might contract prematurely, known as a premature atrial or ventricular contraction (PAC or PVC). Premature contractions are fairly normal and happen to all of us every once in a while. If you've ever felt like your heart skipped a beat, usually in response to caffeine or stress, it's likely that either the atria or the ventricles contracted prematurely. In these rare instances, the SA node will resume its rhythmic beating momentarily and all will be well. PVC may also happen if we are overstimulated from caffeine or an excess of stress hormones such as adrenaline. If this happens regularly, the ventricles may begin to fire early. Then, when the actual signal to beat comes from the AV node, the beats clash prematurely. On an electrocardiogram (EKG), this will look like a series of normal beats followed by a big, wide beat signaling a PVC, where the next beat has clashed with a premature contraction, which you can see in Figure 5.X. Often when feeling your pulse, the PVC will appear as almost a blank spot. PVCs are electrical collisions that show up big on an EKG and small at the pulse. Sometimes they feel like a thud or a gurgle on the left side of the chest. It's great when you can't feel them, though, as they can make you nervous or anxious.

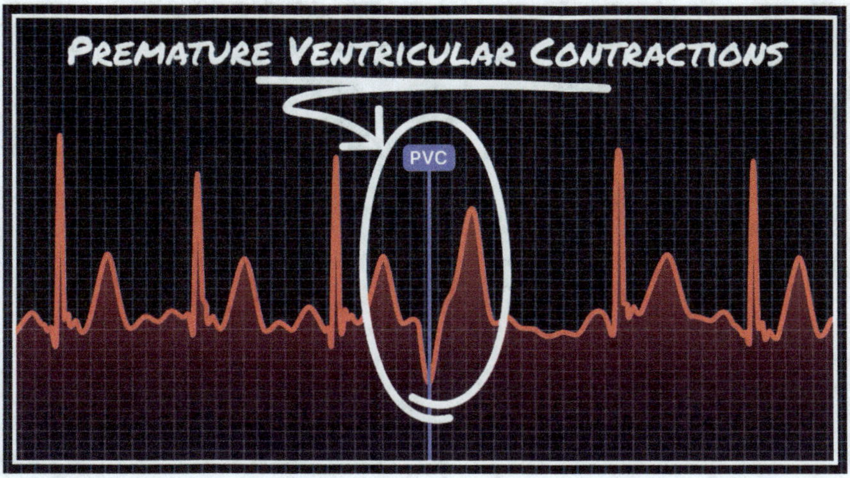

Figure 5.1

In most cases, after a PVC, the following beat will be normal. Occasionally, a PVC will be followed by another skipped beat—an issue called bigeminy. If the clash happens on every third beat, it's called trigeminy; every fourth beat would be quadrigeminy. If not corrected quickly, PVCs may degenerate into ventricular tachycardia and rarely into ventricular fibrillation, in which the ventricles quiver with the rapid electrical pulse and stop altogether. At that point, the only solution is to use a defibrillator to shock the heart back into a normal rhythm. This is somewhat like atrial fibrillation, but far less common.

Sometimes the atria misfire due to overstimulation—called a premature atrial contraction (PAC), similar to PVC. Rather than the beats clashing together, the SA node will send out two beats back to back, as shown in the EKG reading in Figure 5.2:

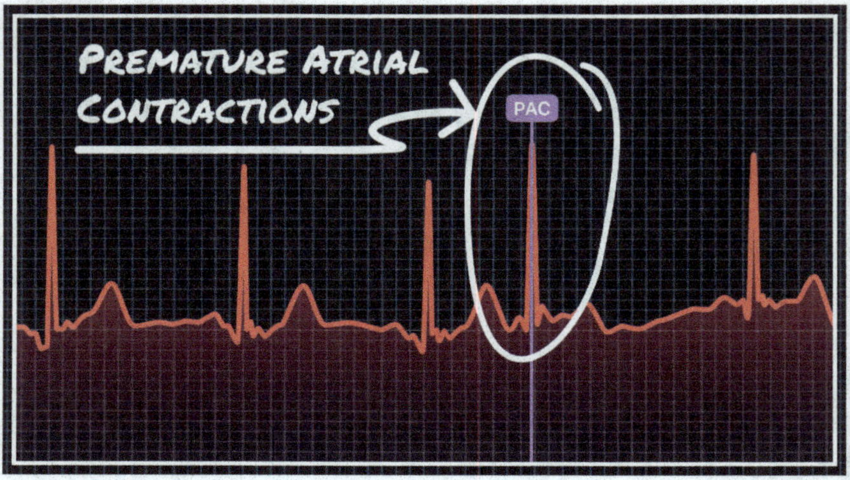

Figure 5.2

If this problem worsens, the beats will stack up on each other in rapid succession, firing like a machine gun, in a condition known as supraventricular tachycardia (SVT), which can occasionally degenerate into atrial fibrillation, in which the firing is not rapid and synchronized beating and more like a Geiger counter.

Sinus tachycardia occurs when the heart beats more quickly than it should, at over a hundred beats per minute at rest. It may occur as a result of drug use, anxiety, or atherosclerosis, or, as you've seen, because of another unaddressed arrhythmia. This faster pulse is not dangerous and often easily corrected with lifestyle changes.

On the other end of the spectrum, the heart may also beat more slowly than it should (under sixty beats per minute at rest) in a condition called bradycardia. Bradycardia often occurs as we age, or in response to a disease or medication, and often causes feelings of lightheadedness and fatigue. Aging can cause a degenerative sickness of the SA and/or AV nodes.

In Chapter 4, you were introduced to atrial and ventricular fibrillations, which occur when the muscle quivers or vibrates, often after the muscle has been pathologically thickened due to genetics, aging, extreme exercise, disease, or high blood pressure. Ventricular fibrillation is, in most cases, swift and fatal. If immediate action isn't taken after the onset of VFib, a patient will almost certainly die of sudden cardiac arrest.

Atrial fibrillation, on the other hand, can be moderated and managed throughout life. In fact, atrial fibrillation is the most common arrhythmia and is becoming a more common problem for Americans. In 2015, data from the Framingham Heart Study revealed that the prevalence of atrial fibrillation (AFib, or AF) has more than quadrupled for men and more than tripled for women in the last fifty years. However, Dr. Anne M. Gillis notes that "because women live longer and AF prevalence increases with age, the absolute number of women with AF exceeds that of men."

The increase in cases of AFib or AV can be attributed to several causes: Body size correlates with AFib, so as obesity increases, AFib does as well. Alcohol consumption also contributes significantly to AFib. Aging does, as well, and we have a huge population of aging baby boomers. A good analogy is that an older house will more often have flickering lights than a newer one.

Patients with AFib are twice as likely to experience heart attacks, and four to five times as likely to experience strokes. Women with AFib in particular have an 18 percent higher likelihood of experiencing a stroke than men with AFib. Besides Kareem Abdul-Jabaar, many celebrities, politicians, and former athletes suffer from this condition, including NBA Hall of Famer Jerry West, show host Howie Mandel, and former president George H. W. Bush. Thanks to the advocacy of these and many other public figures, awareness is growing around AFib and other arrhythmias—though we still have a long way to go

One patient of ours is the CEO of a chain of medical practices in Canada. He works hard and, on top of that, is a fanatical mountain biker. When he came in to see us, I suspected immediately that the strain of his stress and workouts might be having an effect on his heart. On the treadmill test, he lasted forever, but we found concerning irregularities in his rhythm: bigeminy, trigeminy, couplets, and triplets of PVCs. We prescribed him a drug to help address the PVC and worked on optimizing his inflammation and oxidation. A year later, he came back and we did the treadmill test again. This time, his heartbeat was exactly what we hoped to see.

The good news about electrical issues is that most of us can avoid them with the right diet, correct medication, regular (moderate—not extreme) exercise, low stress, and generally staying healthy and avoiding or treating disease. That's not the case for everyone, though. Amazing new ablation procedures have been incredibly helpful surgical approaches for those with resistant AFib or PVCs.

Congenital Conditions

One young patient of ours, Paxton, was an athlete who had been suffering from a strange problem. Every time he played soccer, he would suddenly pass out, much like Clara. Eventually, the problem became so acute that Paxton stopped playing sports altogether and almost decided not to go to college. Through our nonprofit foundation, EveryHeart, we tested Paxton at his high school and discovered that he had an accessory bypass track, a condition known as Lown-Ganong-Levine (LGL) for the physicians who discovered it. Another common form of this condition is called Wolff-Parkinson-White (WPW).

The electrical pathway can be thought of like a railroad track, traveling between the SA and the AV nodes and out through the bundle of His, where it splits into the right and left bundles to the Purkinje fibers in the outer walls of the ventricles, as shown in Figure 5.3.

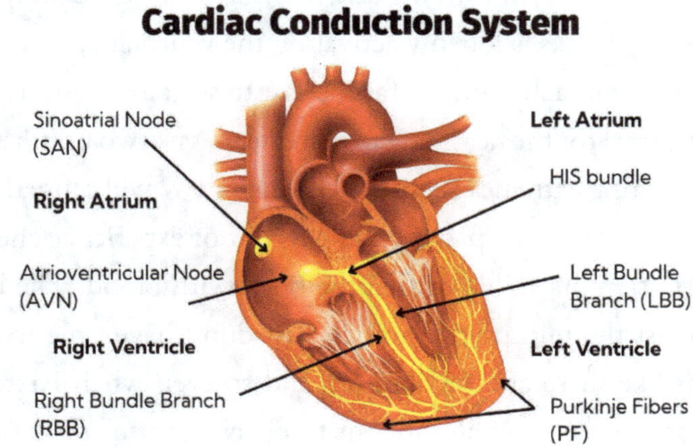

Figure 5.3

In rare cases, children are born with parallel accessory tracks between the SA node and the bundle of His, so that when the heart rate is elevated, as with exercise or stress, the electrical current can jump the track and bypass the AV node altogether, traveling directly from the SA node to the bundle of His, as shown in Figure 5.4. This causes a premature ventricular contraction, accelerating the heartbeat well above a healthy range and causing the patient to pass out. In extreme cases, they may even die.

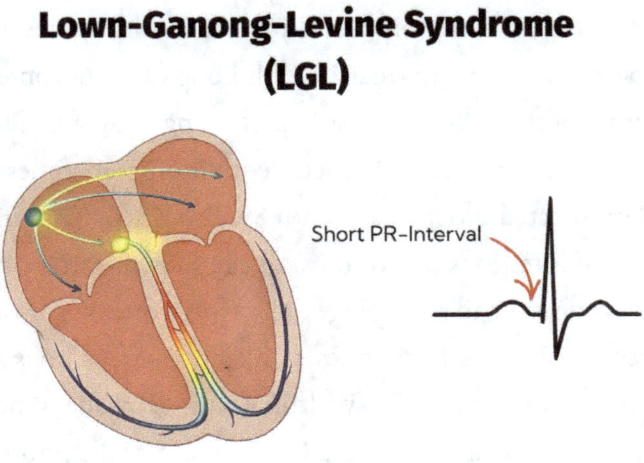

Figure 5.4

When Paxton started working out, his heartbeat would jump to one of his accessory tracks, suddenly activating the ventricles and causing a rapid heartbeat, or tachycardia—far too fast to sustain physical activity. While it's normal for the heart rate to increase to, say, two hundred beats per minute during extreme exercise, LGL and WPW will cause the heart to beat so rapidly that the patient will pass out or experience chest pain as the heart attempts vainly to fuel the body with blood. The heart is beating so fast the pumping chambers just don't have time to fill up. Conditions like these are uncommon, and so even when Paxton had been sick or injured, no one thought to check for it or detect its rare occurrence. Mortality is unlikely, although in some severe cases it may cause sudden cardiac death. Thankfully, the treatment for both conditions is amazingly simple and permanent: a cardiac electrophysiologist can perform a cardiac ablation, repairing the tracks and guiding the pathway back into a normal, healthy rhythm. Paxton was able to go to college and play soccer after all.

Another congenital electrical condition is called Long QT syndrome, in which diastole, the heart's rest period after the beat, is prolonged, causing a variety of arrhythmias. Long QT syndrome is caused by a genetic mutation and often goes unnoticed and undiagnosed until the patient experiences a symptom, usually fainting during stress or exercise. It's possible to live a fairly normal life with Long QT syndrome, as long as you are aware of the risks and avoid exacerbating the issue. On the other hand, Brugada syndrome is a genetic condition sometimes known as Sudden Unexpected Nocturnal Death Syndrome. As you might guess, this condition is responsible for most of the sudden cardiac deaths that occur during sleep, at an average age of just forty-one. It's most common in Southeast Asia, and four times as likely to affect boys than girls. Fortunately, both Long QT syndrome and Brugada syndrome are extremely rare, occurring in less than .01 percent of the population.

Lights Out: Sudden Cardiac Death

On Monday, January 2, 2023, tensions and excitement were high at Paycor Stadium, where the Cincinnati Bengals faced off against the visiting Buffalo Bills. Nine minutes into the game, Bengals quarterback Joe Burrow threw the ball to wide receiver Tee Higgins at the 42-yard line. Higgins ran as the Bills closed in around him, colliding with the Bills' safety, Damar Hamlin, who rushed to tackle him at the 50-yard line. The two players fell to the ground in a successful tackle. But as they stood, it became immediately clear that something wasn't right. Hamlin took two steps backward and fell limp on the field, in what officials and medical personnel quickly realized was some form of cardiac arrest. For nineteen minutes, the players on both teams, and millions of viewers around the globe, watched in horror as medical personnel administered CPR and used a defibrillator to try to shock his heart back into rhythm. The stadium was eerily silent and ESPN cut to commercials while the world waited for the outcome. The players knelt and bowed their heads in prayer. Thankfully, their efforts worked; Damar Hamlin's heart restarted. The stadium erupted in cheers as the ambulance carried him off of the field—in critical condition, but alive.

With every beat of the heart, the electric pulse is passed from cell to cell throughout the electrically charged muscle in a wave that happens in mere milliseconds. Normally, electricity travels at the speed of light, but because the charge must be conducted from cell to cell, the movement here is a bit slower—a detail which is critically important in explaining what happened to Damar Hamlin. After a heartbeat, the cells are at complete rest for about twenty milliseconds before the SA node begins the next beat. When doctors are getting a CT scan of the heart, it has to be timed in that brief twenty milliseconds in order to work—which is quite difficult to accomplish, as the heart is almost constantly rocking back and forth.

But in that split second of time, a moment that even doctors struggle to capture with the most advanced medical technology that has ever existed, Tee Higgins and Damar Hamlin collided, the blunt force trauma completely stopping Hamlin's otherwise healthy heart. This event is called commotio cordis, also known as a cardiac concussion, and is fatal in about 65 percent of cases. It's rare for commotio cordis to occur in adults, even pro athletes, although the chances of commotio cordis double when the heart rate exceeds 120 beats per minute. More commonly, though, it affects young teens while they're playing sports, likely because their hearts and rib cages are still developing. Many youth sports teams, especially baseball or lacrosse, where the chances of a stray ball hitting someone's chest are much higher, require participants to wear chest protectors—although the efficacy of this gear has been hotly debated.

It's estimated that sudden cardiac death (SCD) accounts for over thirty thousand deaths in the United States every year, the vast majority of them (about 90 percent) occurring after strenuous exercise. It's easy to miss signs that a more serious issue is imminent, like shortness of breath and dizziness, which are also common symptoms of working out.

Although commotio cordis is an electric issue, any form of heart disease can be a factor in sudden cardiac death. The National Institutes of Health define sudden cardiac death as "death due to a cardiovascular cause that occurs within one hour of the onset of symptoms. A sudden cardiac arrest occurs when the heart stops beating or is not beating sufficiently to maintain perfusion and life." SCD might be caused by a myocardial infarction, a ruptured aortic valve, cardiomyopathy followed by ventricular fibrillation, stroke, pulmonary embolism, or any of the other cardiac events we've discussed in the previous chapters. The four types of heart disease are also interrelated, one often leading to another in an intricate dance of damage that can wreak havoc in our lives. We see this stark truth with the NFL teams we've worked with throughout the years.

In 1986, Denver Broncos special teams coach Mike Nolan lived in our neighborhood, and we were friends with him and his wife Kathy. He came to me for testing and was so impressed by what we did that he told head coach Dan Reeves about us. Reeves had been a star in his own right as a player for the Dallas Cowboys, helping them to two Superbowl victories in the early seventies. One of Reeves's best moves as coach was empowering quarterback John Elway, who would eventually lead the franchise to two Super Bowl victories before his retirement. But the next best thing Reeves did was to work with us to detect and prevent potentially fatal cardiac events for the Denver Broncos.

After my time working at the Cooper Clinic came to an end, I returned to operating my own clinic in Denver, somewhat more sustainably this time, as I was operating independently with my dream of stopping heart disease still intact. I was doing treadmills and EKGs, but I needed a CT scanner in Denver because I was still sending patients back to Dallas for coronary calcium scoring. Eventually, an entrepreneurial anesthesiologist named James Erlich, recognized the potential of what we were doing and brought in a bunch of investors to found Colorado Heart Imaging and buy an electron beam CT scanner. No doubt, I was the happiest person about this development. Further, CHI had a beautiful room where I could do consultations with patients, including many of my notable patients with the Denver Broncos.

Through Dan, we began working with Steve "Greek" Antonopoulos, the most famous trainer in the history of the NFL. Steve was the thread between different generations of Broncos. Coaches might be fired, but the training staff was always there watching over generations of coaches and players as they cycled through Denver. We started off with treadmill testing for the coaches, but with the new CT scanner available, I soon persuaded Steve and Dan that we should add calcium testing as well, to identify the coaches most at risk. The next year, Steve wanted us to start testing the players. Every year, we did more and more testing: coaches, players, and all the rest of the staff of the Denver Broncos. We found that the players most often die from problems with the muscles or electricity, like Damar Hamlin, but coaches were most at risk of dying from atherosclerosis, due to the combined effects of hypertension (high blood pressure), genetics, aging, and cholesterol.

The success of the Denver Broncos program, shared by Steve throughout the rest of the NFL, soon led to testing all future NFL draftees at the annual NFL Combine in Indianapolis to identify potential risks for players even before they were drafted to major teams.

By the mid-nineties, we spotted another vulnerable group: the alumni. These professionals were no longer in danger of death from the contact of the sport, but they were still at risk of atherosclerosis, arrhythmia, and heart failures. Now retired from the NFL, most of these men had lost access to the testing and resources we were providing to current players. So we started the Denver Broncos Cardiovascular Prevention Program. Not one dollar was received by anyone involved with this pioneering program. However, we were able to inspire many corporations to donate free medications, EBCT, calcium scores, echocardiography, lab profiling, marketing, and treadmill testing. Literally millions of dollars' worth of non-cash support was given to help former players and raise awareness about early detection and prevention of heart disease. We highlighted the dangerous reality that former athletes were at heightened risk for premature heart disease. Everyone was beginning to realize the risks of dementia and traumatic brain injuries from playing football, but for the first time, people started to recognize this much more common and fatal danger as well. All of a sudden, the NFL began to get bad press about so many of their former athletes falling prey to heart disease without enough support from the league.

By the early 2000s, I found myself at the NFL offices on Park Avenue in New York City, across from the Waldorf Astoria. The lobby looked like a scene out of a movie, with stark white walls, one giant TV screen, and every single Superbowl ring on display, from the very first to the last. Behind these walls were offices where I knew dozens of people were working. I was brought in to present a potential program for all alumni to the NFL medical committee, about thirty austere doctors. These were the East Coast elite, the best doctors from places like Harvard and Columbia. They didn't quite know what to make of me; they probably thought anyone from west of the Hudson River had come in on a horse. I decided to give them my normal lecture, with all the passion and excitement I could share about what we had been doing with the Broncos. I knew they needed something truly unique and cutting edge in order to get past the bad press. I told them that if they were to start this program, they shouldn't just provide the bare minimum standard of care, which most ex-players could get down the street at their doctor's office. Many of their alumni were already doing just that, and it wasn't working. Nobody would stop heart disease with the bare minimum of treatment. The average doctors weren't paying much attention to advanced cardiac imaging. I told the committee that we needed to provide the best, most extensive, state-of-the-art testing for their alumni. There wasn't a single smile in the room. I left that day feeling unsteady, but hoping that something I had said had made an impression.

Days later, I got the news: the NFL had decided to fund our program. We were off to the races, flying all over the country and persuading facilities to donate as many resources as they could to help with the cause. Occasionally, a player would say something like, "It's about time the NFL started taking care of us" in a local paper, and I'd receive a call from the Commissioner's office threatening to pull our funding if the players were ungrateful. With the help of my growing team, I was able to walk the delicate line between the league and its ailing, sometimes combative former players. We continued to operate this remarkable nationwide program for about five years from our little office in Denver, saving hundreds of alumni and raising awareness of heart disease across the country.

Eventually, the Cleveland Clinic and the Mayo Clinic emerged as the new centers for NFL Alumni care, taking over our program but, thankfully, leaving our treatment philosophies intact. In the years since, the program we created has expanded beyond just the prevention of heart disease to assist retired players with improving their quality of life in many facets, from traumatic brain injuries to prostate care to mental health awareness, and lives on as the NFL Player Care Foundation, created in 2007. From my perspective, the NFL program was a vital stepping stone to the Boone Heart Institute's mission to eradicate heart disease and stroke.

The Match

While we've discussed many of the contributing factors that predispose us to all forms of heart disease, there is one catalyst for all of them that we've only just touched on.

On December 30, 2021, around 11:25 a.m., what appeared to be a small brush fire broke out along Marshall Road in Superior, Colorado. At first, all anyone could see was the smoke clouding the residential roads, but as the winds picked up and a small shed spontaneously burst into flames, residents and authorities began to realize that a far more serious problem was brewing. By noon, the fire were blazing out of control, raining ash over the streets and devouring homes and stores in its path. The neighborhoods in Superior are filled with beautiful homes, but the trees around them were matchsticks waiting to be ignited. Several of my patients' homes were only one house away from being burned. The streets of Denver, thirty miles away, looked like a war scene, with the sky red and smoke clouds billowing. The whole city was covered in an inch of soot. By the time the Marshall Fire was extinguished by heavy snowfall on December 31, it had burned over six thousand acres, razed at least a thousand homes, and killed two people and innumerable pets—the most devastating natural disaster in recorded history for Boulder County.

Deadly wildfires like this often appear to come out of nowhere, but there are a number of precipitating factors that set the stage for disaster. In the spring of 2021, we had had an unusually wet winter and spring, with months of rain and snowfall so thick that in certain areas people couldn't even get to their houses. The result was a beautiful, lush summer where the brush and grass grew thick and green over the hillsides. Unfortunately, the summer months that followed were incredibly hot and dry, killing all the greenery, and turning it into a tinder pile just waiting for a match.

In fact, there were two matches. The investigative report released by the Boulder County District Attorney's office revealed that, nearly a week earlier, residents in the area had started a controlled burn to eliminate old pallets and debris. Firefighters had been called out to the area and verified at the time that everything was safe and legal. Unfortunately, by the time the extreme winds began on December 30, there were still hot remnants of the fire left over, and the blaze quickly reignited. Not only that, but just two miles down the road, "hot particles" from the powerlines ignited another fire at the Marshall Mesa trailhead one hour later. The match was lit, and disaster ensued.

I often think of heart disease like those dry, grassy fields and the tinder trees—structural defects, bad genes, accessory electrical tracks, high counts of apoB-bearing LDLs, and plaque accumulating in the arteries—any one of them, and likely several of them, providing the necessary fuel for a sudden cardiac death. All that's needed is a single match. For some elite athletes, the match might be extreme exercise. But, again, most of us would be better served by exercising *more*. No, for most of us, the match is something far more common, something you likely battle on a weekly if not daily basis.

Stress.

CHAPTER 6

The Fatal Cost of Stress

On a warm August day, a sixty-year-old woman was driving with her husband, headed home to Denver from a weekend in the mountains. She was the picture of health—and I should know. She was my patient.

Suddenly, though, she looked at her husband with an odd expression, and steered the car toward the shoulder of the road. She slumped over the wheel and her husband, panicking, dragged her out of the car. Her heart had stopped completely. Frantic, he started performing CPR. In a stroke of divine intervention, a passing policeman and a nurse both stopped to help. They called an emergency medical team, but they were still too far from Denver for a ground ambulance to get there quickly. A helicopter was dispatched to evacuate her to the nearest hospital and hopefully get her enough oxygen to keep her alive in the meantime.

When they arrived at the hospital, her husband called me to let me know and to ask for advice. Together, we wondered: how could this have happened? We'd cleaned out her arteries; she had no evidence of atherosclerotic plaque anywhere. She was already taking Repatha to lower her overall LDL count. Her treadmill tests were superb. There were no electrical problems, and her blood tests came back negative for any of the genetic markers of heart disease.

At the hospital, there was a tense moment where, technically, she died, and her husband wondered if all was lost. Thankfully, the medical team worked tirelessly and succeeded in bringing her back to life. But for the next two days, she remained in a coma, everyone around her unsure if, ultimately, she was going to make it.

I was completely mystified. We had done everything right. How could this be happening?

As we spoke over the phone in those days, waiting to see if she would come through, her husband told me of the immense stress that his wife had been under. Their adult son, who had psychiatric problems, had been living with them and wreaking havoc in their lives. He had been seeing a psychiatrist, but the problems he faced weren't improving. The situation in their house had become untenable; the parents took a weekend away to the mountains to relax, get some clarity in the mountain air, and come up with a plan. There, they made the heart-wrenching decision to kick him out.

But his mother wasn't at peace with it. She saw the necessity of the decision, and yet, it broke her heart.

Literally.

There's a condition that was first described in Japan called takotsubo cardiomyopathy. Takotsubo is a type of Japanese pot used by fishermen to catch octopuses. The pot has a narrow opening that leads into a wide bowl, making it easy for the octopus to climb in, and nearly impossible for it to climb out. When someone, usually a postmenopausal woman, suffers from takotsubo cardiomyopathy, the bottom of the left ventricle balloons out, creating a resemblance to the unique Japanese pot. In this condition, the left ventricle, the hardest-working chamber in the hardest-working organ, is unable to squeeze blood into the aorta and out to the rest of the body. Today, this condition is often called a stress cardiomyopathy, and it is most commonly triggered in times of intense psychological or emotional stress. The patient suffers from a broken heart.

My patient's husband believed that after they'd spent a wonderful weekend together in the mountains, the stress of returning home to enforce the hard decision they'd made continued to mount, until her heart couldn't take it anymore. She couldn't face returning to their reality and dealing with their son.

Thankfully, the medical team was able to put in a pacemaker, and the wife started to come back to life. But when the husband told their son about his mother's condition and her near miss with death, he didn't seem concerned about her. He certainly wasn't going to change his ways or try to make life easier for his parents. The husband knew that in order for his wife to make a full recovery, they had to get him out of their house before they returned home. He called the police to have his son removed, a difficult decision made easier by the necessity of prioritizing his wife's health. In the next week, his wife was discharged from the hospital, and they were able to return home to begin recovering, both physically and emotionally, from their trauma.

For me, this was a stark reminder that no matter how healthy the heart is, stress is the ultimate risk factor. My patient had everything going for her—no clotting, no thrombosis, no atherosclerosis, and no plaque. She was taking all the right supplements and medications. And yet, a broken heart nearly took her life.

You've probably noticed already from the stories in this book that physical stress is the most common trigger for heart disease and acute cardiovascular events, like a heart attack. Mental stress is just as dangerous. In Chapter 4, we discussed the impact of extreme *physical* stress caused by excessive exercise and specific physiological factors like the size of the heart. But in this chapter, we'll dive further into the ramifications of psychological and emotional stress on our hearts. The link between mental and emotional stress (both acute and chronic) and heart disease is so profound that it can no longer be ignored. Despite the evidence staring at us, many people don't understand the fatal risks of living in a state of chronic mental and/or emotional stress. The impact of this stress can be so severe that even if the heart is healthy, the right amount of stress can cause a cascade of disease that can send even the strongest heart to an early grave.

Testing for Mental Stress

The link between stress and heart disease has long been suspected, but Kenneth Cooper was the first physician to start systematically using the treadmill test to judge his patients' cardiovascular health and identify those at risk for cardiovascular disease. This is an excellent diagnostic tool, and one that cardiologists and doctors of preventive medicine, including myself, continue to use. When combined with an EKG, the treadmill test gives us invaluable information on your heart's efficiency, strength, and recovery time when put under strain.

The only problem with it is that it's a measure of the test-taker's ability to cope with *physical* stress. Like my patient who suffered from takotsubo, many people in top physical shape can pass a treadmill test with flying colors, and yet still succumb to the impact of psychological and emotional stress on their system. In fact, as we discussed with exercise and the J-shaped curve in Chapter 4, regular and moderate amounts of physical exercise have a very positive effect on the body. It's only when that physical stress becomes excessive and even *obsessive* (as with obligate runners like Jim Fixx) that it can have devastatingly harmful effects on the cardiovascular system. Mental stress, on the other hand, always has a negative impact on the body, no matter how big, small, or prolonged that stress is. But until recently, we didn't have methods for identifying who is most sensitive to the impact of mental stress on their system.

I started to become attuned to this gap while working with Dr. Robert Eliot, founder of the National Center of Preventive and Stress Medicine in Phoenix, Arizona. Although at the time, mental stress was not considered part of the pathology of cardiovascular events, Eliot himself had experienced a heart attack earlier in his career, and he believed stress to be the catalyst. He'd also noticed in his work with NASA that otherwise heart-healthy astronauts experienced worrying electrical patterns after simulated takeoffs, putting them at real risk of a major adverse cardiac event in space. His book, *Is it Worth Dying For?*, raised awareness in the general public around the evils of stress, but it didn't garner a lot of support within the medical community. He needed the numbers—the quantifiable proof that stress played a role in heart disease.

There are over ten thousand neuropeptides, or signaling neurons, in the brain involved in the stress response, making it incredibly difficult to measure in the moment. So at his practice in Phoenix, we experimented with other ways to measure it, particularly through changes in blood pressure and heart rate. We were looking for "hot reactors," Eliot's term for relatively healthy people who tend to experience outsized physiological and emotional responses to triggers, with innovative new tests designed to measure the efficiency, strength, and recovery times of the heart while under mental, rather than just physical, stress.

Eliot designed a stress-testing machine kit that came in a box, which he called the Portable Stress Laboratory. He aimed to sell it to other physicians to help them identify the hot reactors among their patients. It was about the size of an old box TV, and cost about $10,000 per machine. The kit included a sophisticated blood-pressure tracking system that would automatically record the data, and a simple ping-pong video game that patients would play while having their blood pressure monitored. It also instructed patients to perform simple mathematics and other mental tests. It was a great idea, but it never caught on. It was too expensive, and the kit was too clunky and labor-intensive for most physicians' offices. After I left Eliot's practice, he gave me a bunch of the kits and they sat in our garage for years. My sons had lots of fun playing the ping-pong game on them while growing up, but that was the extent of it.

While in Phoenix, I was Eliot's right-hand man, and I took on the primary role of working directly with our patients. Eliot preferred to be on a stage, writing books, or traveling the world promoting his story and ideas. But in working with patients every day, and later after starting my own practice, I saw that there were much more accessible and effective ways to test for mental stress without the requirement of a large, cumbersome, and costly machine. The game might be fun, but I could teach doctors how to measure the same data using equipment they already had in their offices, like blood pressure machines and heart rate monitors, which are ubiquitous in medical offices—no game required. I found I could sell the philosophy and the methods, without charging physicians a dime.

I developed a systematic protocol for testing mental stress, which I called the Mental Office-Based Stress Testing (MOST) Protocol, as a standard diagnostic procedure for all of our patients. We quickly found that not every person who experiences significant mental stress (and whose body reacts accordingly) expresses that stress outwardly in a "hot" reaction like anger or frustration, as Dr. Eliot had suggested. Many people who experience extreme mental stress keep a stiff upper lip and tend not to talk about it. These people, who appear to be cool as cucumbers on the surface, may experience the most surprising cardiac events, because no one around them can see it coming. In fact, it seemed that mental stress affected nearly everyone.

The MOST (Mental Office-Based Stress Testing) Protocol

The MOST Protocol is a fairly simple series of diagnostic measurements that can be completed in any primary care clinic or office. It has become a cornerstone of the diagnostic assessments we conduct as part of the Boone Heart Protocol (the topic of Chapter 9). As I first described it in a 2003 article cowritten with Dr. Jeffrey Anthony:

"This protocol includes the following components: (1) baseline assessment in which blood pressure level and heart rate are measured at rest in the sitting position; (2) relaxation assessment (deep-breathing relaxation); and (3) stress assessment."

In the stress assessment portion of the protocol, we introduced three types of tests: First, patients are asked to subtract by seven starting with 777 (for example: 777, 770, 763, 756, etc.) while their blood pressure is monitored. Second, patients put their hand into ice water up to the wrist for sixty seconds (if they can make it that long). Finally, patients perform an isometric test, holding a handgrip at thirty percent of their maximum strength for sixty seconds.

The math portion of the test alone turned out to be a revelation. Soon after we started conducting the MOST Protocol, a young, fit engineer came to see us. As expected, he performed extremely well on the treadmill test. Next, we asked him to count backwards by sevens starting at 777. He was confident—almost excited—to do this portion of the test, and raced through the numbers backwards, getting all the way to zero in less than two minutes. It was an impressive feat, but his blood pressure skyrocketed. We already knew from imaging that plaque was building in his arteries. It was a signal to us that under the right stressful circumstances, he was just as likely to be the victim of a heart attack as someone who could barely run a mile.

On the other hand, over the years, we've had patients who are terrible at subtracting and can barely make it past 770, but it doesn't faze them at all; their blood pressure stays constant. Of course, not all patients are excited to take this portion of the test. Once, when one of our technicians was going through the protocol with a famous NFL linebacker, she asked him to start counting for the math test. He looked at her sullenly and just said, "No." I used to joke sometimes that when we were testing the Oakland Raiders, we had to ask them to subtract serially by ones instead of by sevens.

Due to our work with the MOST Protocol, I started getting requests to speak all over the world. Cardiologists in Argentina and Italy were particularly interested in what we were learning about the impact of mental stress. But not everyone was receptive to it; many U.S.-based cardiologists thought what we were doing was too mysterious, almost "woo-woo." This was long before the benefits of mindfulness, yoga, prayer, and other meditative and spiritual practices were widely recognized, and what we were doing perhaps didn't feel "scientific" enough for many doctors. Other patients, particularly businessmen, didn't like the term "stress," and so we took to calling it "mental intensity," which tended to be better received by patients. It took another decade before independent research and continued reports from the medical community on the harmful effects of mental stress allowed those effects to be recognized by society at large.

In the meantime, we continued working with those who were interested in seeing that the MOST Protocol saved lives. One group that hired us early on was the South Metro Fire Rescue District (SMFR), the second-largest fire department in Colorado. Studies have shown that CVD is the most frequent cause of death for firefighters while on duty—not smoke inhalation or burns, as one might imagine. SMFR has been a strong supporter of our work, and by now we've used the MOST Protocol to test nearly a thousand firefighters over the years. We've found that firefighters are a hundred times more likely to die during an alarm response than while asleep. In any given year, if we tested about two hundred firefighters, we would see that about a hundred had high blood pressure, though only thirty or so had scans revealing plaque in their arteries. For the firefighters who had high blood pressure but no plaque, we recommended lifestyle and stress management coaching to help them manage their bodies' physiological response to the alarms. But for those who had both high blood pressure and plaque (especially those whose hs-CRP and Lp-PLA2 tests also revealed hot, vulnerable plaque), we were able to alert the fire department to the ten or so firefighters who were most likely to experience a devastating acute cardiovascular event while on the job. And crucially, we were able to implement aggressive preventive treatments to ensure that those vulnerable firefighters were able to stay on the job. Indeed, we have never recommended retirement or desk duty to any of our firefighter patients.

What Makes Us Stressed?

With many of our patients, we've found that work is the most prominent source of mental stress, and this is confirmed in the research. For example, a longitudinal study of nurses in Denmark found a "1.4-fold increased risk" of heart disease in nurses who reported high levels of work stress. In 2016, a meta-analysis of studies assessing the link between job strain and incidents of ischemic heart disease (IHD) also found "moderately strong evidence" that both men and women who experience job strain also experience higher rates of IHD.

The Japanese have a term for work-related deaths called karoshi. This term was popularized in the 1980s when several young businessmen in Japan, who appeared to be in their prime, suddenly and inexplicably died. The cause of these *karoshi* deaths was found to be sudden heart attacks and strokes, triggered by the immense stress these young men faced at work.

Interestingly, also in the 1980s, U.S. doctors noticed a surprising trend in acute cardiovascular events. Most of them occur in the morning, when blood pressure, which naturally fluctuates throughout the day, tends to be highest. Remember our patient Mark from Chapter 1, who experienced a spike of stress one morning before a high-stakes presentation? It turns out incidents like this are often to blame for triggering a heart attack, or myocardial infarction. One study by Muller et al found that most incidents of myocardial infarction occur between 7 and 11 a.m., and another from Goldberg et al saw that, in most cases, symptoms began within an hour of waking up.

Any work or life event that disrupts the body's natural circadian rhythm, such as shift work, taking a red-eye flight, or experiencing jet lag, is also associated with higher rates of CVD, obesity, and diabetes. Even the transition to daylight savings time is known to cause higher incidents of heart attacks. Your chances of an acute cardiovascular event also increase during the holidays and tax season (especially for accountants).

Of course, interpersonal conflict is another common source of stress that may lead to adverse heart conditions, as with my takotsubo patient. One study from the Harvard Medical School found that women's chances of a heart attack increased by 24 percent after one divorce, and as much as 77 percent after a second divorce.

And the longer mental stress is experienced, the more pronounced the risk for heart disease.

Chronic Stress and Trauma

In a landmark 1998 study, now known as the Adverse Childhood Experiences (ACES) Study, a team of researchers led by Dr. Vincent Felitti in San Diego, California, was able to determine a relationship between several childhood experiences and negative health outcomes. The team surveyed over 9,500 patients at Kaiser Permanente, asking whether they had experienced any of the following adverse experiences as a child: "psychological, physical, or emotional abuse; violence against mother; or living with household members who were substance abusers, mentally ill or suicidal, or ever imprisoned." They found that the more ACES a person had experienced, the higher their likelihood of experiencing alcoholism, drug abuse, depression, suicidal thoughts, sexually transmitted diseases, sedentarism, obesity, and a smattering of other health diagnoses, including cancer and ischemic heart disease.

The ACES study had a profound impact on our collective understanding of the relationship between chronic stress, traumatic experiences, and physical health, opening the door to a field of research that has continued to unlock new insights in recent years.

Several studies since then have confirmed the link between ACES and cardiovascular disease particularly, in all cases showing that the higher the number of ACES, the greater a patient's likelihood of having CVD. In her book, *The Deepest Well*, Dr. Nadine Burke Harris writes, "A person with four or more ACES was twice as likely to develop heart disease and cancer and three and a half times as likely to develop chronic obstructive pulmonary disease (COPD) as a person with zero ACEs."

This reality was driven home for me when I learned about the traumatic childhood experiences of one of my patients.

Richard Bard is an incredibly successful serial entrepreneur and corporate executive. He once owned an entire railroad in China. Now, he continues as an advisor at the company he founded, Bard Capital, and is on the board of directors at Penn State. He and his wife own a gorgeous historic manor in Colorado and have remodeled it to a new age of splendor. I just assumed when I met him that he must have been born with a silver spoon in his mouth.

When Richard came to see us, he was sixty-three, and already had multiple blockages of over 90 percent. At that point, it was too late for us to try to melt the plaque with the Boone Heart Protocol; he was one of the only patients of mine to ever need bypass surgery. Although he had some of the markers for heart disease, I wondered how things had gotten *so* very bad for him at such a young age.

That is, until he told me the story of his childhood.

Richard grew up in an impoverished Russian Jewish family. He and his family immigrated to the United States when he was a young boy, and not long after, when he was just eight years old, his mother passed away. His father remarried, but the family was broken after his mother's passing, and he and his brother basically raised themselves in extreme poverty. Despite their hardships, the two brothers succeeded in the business world and became incredibly influential. But even today, Richard obsessively turns out the lights in his mansion to save pennies. When you were once that vulnerable, you always feel that it could all be taken away from you in an instant. In 2001, Richard was honored for his achievements in business with an Ellis Island Medal of Honor, along with other notable recipients including former Yankees catcher, Yogi Berra, and former New York City mayor, Rudy Giuliani.

I respect and admire Richard all the more for what he has conquered, including heart disease. He remains a patient and is healthier now, in his seventies, than he has been for decades. In the autobiography he wrote, he included a nice mention of me, saying, "Thanks to Dr. Boone, I'm still alive."

Besides adverse childhood experiences, other occurrences of chronic stress also contribute to higher incidents of CVD. For example, living in an area with high noise pollution (such as sirens, loud traffic, or other consistent noises, particularly at night), has a positive correlation with hypertension, myocardial infarction, and stroke. Having a low household income and experiencing incarceration are also both associated with higher instances of CVD and myocardial infarctions.

These problems disproportionately impact people of color, immigrants, and minorities across the United States, shedding further light on the impact of systemic racism on healthcare. A 2020 Presidential Advisory from the American Heart Association raised the alarm about the role of systemic racism as a contributing factor toward disparities in heart disease. The authors of the advisory noted that "Black Americans continue to experience the highest mortality rates attributable to CVD and stroke, with almost 30 percent higher CVD mortality and 45 percent higher stroke mortality than non-Hispanic White Americans.**16**"

It's clear that mental stress has a drastic impact on the cardiovascular system, constituting, in most cases, at least an increased risk factor for heart disease, if not also triggering an acute cardiovascular event. But why? Let's take a deeper look at how mental stress changes our physiology, particularly with regards to the cardiovascular system.

The Perfect Storm: Mental Stress and the Cardiovascular System

In his 2014 groundbreaking book *The Body Keeps the Score*, Dr. Bessel van der Kolk wrote, "After trauma, the world is experienced with a different nervous system." That is, stress, especially chronic, prolonged, and unresolved stress, permanently impacts our body's health.

When we experience mental stress, the body interprets that stress as *danger* and activates the sympathetic nervous system. This is a biological defense mechanism designed to protect us in the days of hunting and gathering, such as when Daniel Boone might have crossed paths with a baby bear and its overprotective mama. As soon as danger is sensed, the amygdala, located in the limbic system in the brain, sends a signal to the hypothalamus, which in turn alerts the pituitary glands and the adrenals to release cortisol, commonly called the stress hormone.

The release of cortisol is like a flag, signaling just about every system in the body to either shut down various functions or ramp them up, based on what the body thinks we need in the moment to survive. This happens involuntarily, without our even being conscious of it, in response to any danger, whether that danger is a bear or an impending divorce. Within the cardiovascular system specifically, ten key changes occur that create the perfect storm of atherosclerotic factors.

First, our heart rate increases (1) in order to pump blood more quickly in case we need to run away from the danger. In some people, the stress may even cause a tachycardia or arrhythmia. And as we've learned, when the heart works harder, the muscle thickens, which is called hypertrophy (2). Stress also has a lusitropic effect (3), causing the heart not only to thicken but also to stiffen and become less flexible and more rigid.

As the heart is racing to get blood out to the body more quickly, our blood pressure surges (4). In a relatively healthy person, when we conduct the MOST protocol, we expect to see their systolic blood pressure (the top number in a blood pressure reading, which measures the strength of the left ventricle squeezing blood out into the aorta) rise, but their diastolic blood pressure (the bottom number, which measures the blood pressure when the heart is at rest between squeezes) stay relatively low. For example, on a treadmill test, our eager engineer's blood pressure may have been 180 over 72. However, if a person experiences repeated or chronic stress, the heart has very little time to rest and so that bottom number might be much higher, maybe 85. We see the same effect in patients with hypertension, or high blood pressure. This results in less oxygen going to the lungs, which is why someone under stress might have trouble breathing or hyperventilate.

The heightened blood pressure throughout the system also increases the amount of sheer stress on the lining of the blood vessels (5)—at the same time that those blood vessels are constricting in an effort to preserve oxygen for our most vital organs, called vasoconstriction (6). Imagine a high-speed river blasting through a canyon whose walls are closing in; this is essentially what's happening inside your blood vessels. As you can imagine, this damages the arterial walls, leaving them weak and even more vulnerable to eruptions of plaque.

Further, as the body prepares to face a threat, it increases the blood's coagulability or tendency to clot (7), just in case we need to heal a wound. Simultaneously, stress triggers an imbalance of LDL particles, called dyslipidemia (8), which in turn prompts the body to release macrophages (9), the immune cells that incite an inflammatory response to kill those extra LDL cells and coalesce into foam cells—i.e., plaque in the arteries.

The final kicker is that the kidneys begin holding onto sodium and dumping potassium (10), an essential ingredient for the heart's electrical pathways. With too-low levels of potassium, the patient is in danger of ventricular fibrillation and sudden cardiac death.

When you understand the biology of stress and its impact on the cardiovascular system, it's easy to see why even a seemingly small stressor, especially if someone has been experiencing weeks, months, or years of cumulative stress, may trigger an acute cardiovascular event like a heart attack, or even sudden cardiac death if the heart suddenly stops beating. As if all of this wasn't enough, keep in mind that what I've described here are only the effects of stress on the *cardiovascular system*; stress has further consequences on other systems within the body as well, such as the digestive system.

This is why testing for mental stress is a crucial part of the Boone Heart Protocol. In Chapter 9, we'll explore ways you can mitigate stress and cultivate healthy habits for optimal cardiovascular functioning.

Your heart is depending on it.

Too Little, Too Late

Recently, I took my grandsons to the movie theater to see *Dune: Part Two*. We walked into the lobby, and just in front of the popcorn counter was a team of EMTs and police performing CPR on a middle-aged man who had just experienced a sudden cardiac event. If I had been there earlier, I might have been able to help, but by the time we arrived the emergency medical team had already taken over. The tension in the room was palpable. People crowded around, and eventually the police set up a partition with the cardboard *Star Wars* cutout to keep onlookers from interfering. Unfortunately, the man passed away there in the theater lobby.

Incidents like these remind me every day just how urgent our mission is to stop heart disease. I wonder, did anyone in that man's life know that he was at risk of a heart attack? What triggered the spike of cortisol that proved to be too much for his system? Had he just watched a movie with heart-racing graphics and a booming sound system in the state-of-the-art theater? No doubt he woke up that morning without an inkling that that day would be his last. The Boone Heart Institute was within driving distance. If we'd started seeing him a few months or years ago, perhaps he would have lived another happy thirty or forty years.

But this is just another example of the failure of our medical system to identify those who have heart disease or are at risk of developing it. Cardiologists are often praised, and rightly so, for their skill in performing lifesaving interventions, like cardiac stents, ablations, and even bypass surgeries. But those interventions are only valuable for the patients who can make it to the hospital in time to take advantage of those services. For some, like this man who died a completely preventable death, those interventions are too little, too late.

Wouldn't it be better if we didn't need them at all?

CHAPTER 7

Why Heart Disease Is Winning

Leslie looks at the clock in the conference room, frowning at the time. She and Mark were supposed to discuss the quarterly earnings report and make a game plan for the board presentation this afternoon. She saw him just half an hour ago, as he was walking back to his office with a cup of coffee. It's not like Mark to be late. Where is he?

Remember Mark, our biotech executive from Chapter 1 who experienced a heart attack? Thankfully, his attentive colleague decides to go to his office to see where he might be. Leslie knocks and is concerned to hear a crash in response. She opens the door and sees Mark slumping over his desk, his coffee untouched, struggling to breathe. He looks at her, unable to speak through the pain, panic in his eyes.

Immediately, Leslie helps him sit back in his chair and elevate his legs.

"Mark, don't worry. I'm calling an ambulance for you," she reassures him, dialing 911 on her phone.

With the emergency team on the way, she runs back to her office to grab the bottle of extra-strength aspirin she keeps on hand in a drawer. She hands him four pills and tells him to start chewing. Mark stays conscious and Leslie's reassurances help him to remain relatively calm until the medical team arrives. Still, Leslie asks her assistant to get the defibrillator ready—just in case.

When the ambulance arrives, Mark is loaded onto a stretcher and swiftly transported to the nearest hospital's emergency room, where doctors perform an echocardiogram, angiography, and blood tests to diagnose the heart attack. The coronary angiogram immediately identifies the "widowmaker" blockage in his coronary artery, and the team of doctors moves him to the operating room for an emergency stent placement to restore blood flow to the heart and prevent any lasting damage to the muscle.

Mark is put to sleep, and the cardiologist on staff inserts a catheter into his wrist, guiding it to the blockage in the artery. There, a small balloon at the end of the catheter is inflated to press the plaque down and widen the artery. This procedure is called an angioplasty, and is one of the many lifesaving techniques that cardiologists use to prevent deaths from heart disease. Then, a drug-eluting wire-mesh stent (which is coated in protective medication) is imprinted into the artery wall to keep the affected portion of the artery open.

The surgery is successful; Mark's heart receives the oxygen-rich blood that it so desperately needs and the muscle once again begins pumping powerfully to send blood throughout the rest of his body. He's put on a regimen of blood thinners and statins to prevent another blockage. Thanks to the swift and knowledgeable care of his colleague and the skilled work of his medical team, Mark's life is saved. After the surgery, he's moved to another room to recover, where he is reunited with his wife and daughter. There's not a dry eye in the room.

Over the next day, Mark remains at the hospital for monitoring. He has multiple EKGs to ensure his heart is beating as it should, and the nurses check his blood pressure and other vital signs often. Mark is lucky; he has no other medical conditions that would complicate his recovery, and he is discharged after a brief stay.

He returns home, grateful for the lifesaving treatments he received, but also shaken. This is the kind of pivotal moment that sets the course for the rest of a person's life. He will never be the same. Now that the reality of his disease has been discovered, Mark knows that he's at high risk of it happening again. For the rest of his life, Mark will be on medications. He'll have regular scans and state-of-the-art imaging to monitor the plaque buildup in his blood vessels. After all, if one portion of his network of blood vessels is diseased, it is highly likely that other spots in his brain and elsewhere will also soon need attention. Unlike many less socio-economically advantaged people, Mark is lucky to have access to high-quality medical care and health insurance to pay for these interventions.

But he can't help wondering: Why now? Why couldn't this have been identified sooner?

And what can we do about it?

The good news about heart disease is that we have a professional workforce of highly skilled cardiologists with access to state-of-the-art technologies and lifesaving drugs that can and do rescue people from death's door every single day—if they have access to insurance and healthcare. Too many people don't. But even for those who do, this "good news" is not enough. No one wants to end up on the operating table. As Mark has discovered, the reality of a life with heart disease is one of constant supervision and monitoring, and always with the worry that *if it happened once, it could happen again.* For those who experience one of the life-changing interventions offered by cardiologists and find themselves with a second chance at life, it prompts careful reflection and introspection. What could they have done to prevent this?

Unfortunately, the answer is complex. Our current Western medical system is designed to reward *fixing* disease rather than preventing it. Cardiologists focus on preventing sudden early deaths, and the heroism of their efforts cannot be overstated, but the fact that we need them is a fundamental problem in our society.

In this chapter, we'll discuss the most common treatments for heart disease, which can be summed up as a valiant effort, but often too little, too late. As we've seen, the heart is the center of a complex network of vessels/pipes and electrical currents, or rhythms, pumping our blood throughout the body. The work of cardiologists is akin to that of plumbers and electricians, and the most common treatments for repairing damage are not dissimilar to the fixes for the pipes and valves under your kitchen sink. Essentially, their job is to "clear the pipes" and turn the lights back on. Stents, angioplasty, and valve replacements are all common surgeries that involve addressing the physical malfunctions of our heart parts that help to stave off the impending problems of heart disease, but they do nothing to prevent it from happening again.

Moreover, you can only fix the disease you know about. We'll explore the complex systemic reasons that heart disease goes unnoticed for so long, and unpack how the problem is perpetuated through a dysfunctional medical system that prioritizes reaction over prevention. My hope is that this chapter will convince you to take action to identify any potential warning signs of heart disease in your own life and not settle for the strategy of "wait and see," before something goes truly wrong.

Hail Mary: the Last-Ditch Efforts to Save Your Life

The night of December 28, 1975, was biting cold at the Metropolitan Stadium in Bloomington, Minnesota. It wasn't a problem for Minnesota's home team, the Vikings, the reigning National Football Conference champions. But for the visiting Cowboys, from much warmer Dallas, the wind and chill put them at a distinct disadvantage.

The Vikings took the lead and by minute 59, the Cowboys' defeat seemed all but guaranteed. With the score 14-10 and just thirty seconds remaining, quarterback Roger Staubach took a chance. After the snap, he feinted to the left, clearing the way for wide receiver Drew Pearson to sprint toward the end zone. Sending up a quick Hail Mary prayer, Staubach launched the ball 60 yards downfield. He slightly undershot the pass, but Pearson corrected and reached back for the ball, colliding with Minnesota cornerback Nate Wright. Pearson emerged with the ball firmly in hand and ran it into the endzone for a successful touchdown in what became known as the first official Hail Mary pass. It was a defining moment in football history, and now Staubach's famous prayer has come to describe any long-shot attempt at achieving something monumental, even miraculous.

Much like the lifesaving treatments for heart disease.

As you've already seen, the first symptom of heart disease is often death. For these patients, the incredible skill of cardiologists comes too late. But for patients like Mark and Cameron who do make it to the hospital with emergent heart conditions or even in the midst of cardiac arrest, cardiologists are like quarterbacks, called in for a long-shot, Hail Mary pass. Much of the time, they do this brilliantly. As long as they get there in time, the success rates for surgeries such as coronary artery bypass grafting (CABG), one of the most common and nearly routine surgeries to address heart disease, are over 90 percent. And cardiologists' skills, technology, and research-based knowledge increase every year, with new minimally invasive surgeries debuting in hospitals all the time, leading to increasingly positive outcomes.

At the end of the day, though, it's not until something goes drastically wrong that cardiologists are called in to save the day with one of the following Hail Mary attempts to save your life.

Let's go over the most common surgeries and techniques used to address heart disease.

Angioplasty

As a general rule, cardiologists (and most doctors) prefer to try the least invasive procedures to address a problem first before resorting to more extensive procedures that incur greater risk to the patient and require longer recovery times. Therefore, as a first line of defense against ischemic heart disease, cardiologists may attempt one of a few percutaneous coronary interventions (PCIs), where the focus is on restoring blood flow through the existing, blocked artery using a number of possible strategies. As we've already seen with Mark, one such strategy is an angioplasty. In this minimally invasive procedure, a catheter is threaded into the blocked artery through a vein in the wrist. Then a balloon at the end of the catheter is inflated to widen the diameter of the blood vessel and compress the soft plaque.

In many cases, an angioplasty will not last very long; it's a short-term solution, as the plaque will immediately start building up again.

Stent Placement

The next step, then, is a stent placement. Following an angioplasty, a small mesh tube called a stent may be inserted into the blood vessel in order to cover the volcano of plaque in the artery and prevent the walls from closing in again. Sometimes the use of certain drugs, like statins, which Mark was given, are also prescribed to reduce the risk of restenosis, or the subsequent re-narrowing of the artery.

The problem with stents is that they are made to cover only the blockage that is currently the softest and most vulnerable plaque—the volcano that is erupting or is about to. Using an intravascular ultrasound (IVUS) angiography or Fractional Flow Reserve (FFR), if a cardiologist finds a 70 percent blockage, they will put a stent there at the site of the thickest blockage. However, as we discussed in Chapter 3, plaques tend to extend far beyond that apex, winding throughout the blood vessel. In fact, the most dangerous plaque is not the 70 percent blockage, but a vulnerable plaque upstream or downstream from that block, which might break off and cause an immediate blockage as a clot tries to seal the break. Ideally, stents should be long enough to cover not just the volcano of soft plaque, but all the plaque around it as well.

Unfortunately, this constitutes a logistical problem. In order for them to be placed and at all effective, stents must meet certain design criteria for flexibility, durability, curvature, and lumen diameter. A 2017 assessment of stents by Watson et al found that "currently available stents all reflect a compromise between competing desirable features and have subtle differences in their performance characteristics. Striving to achieve stents with optimal deliverability, conformability, and radial strength led to a reduction in longitudinal strength." In other words, it's nearly impossible for stents to meet all the ideal criteria. As a result, most stents aren't long enough to be completely effective, and they do nothing to stop the continued development of atherosclerosis throughout the rest of the body's sixty thousand miles of blood vessels.

Historically, CABG, angioplasty, and stents were sometimes used to try to prevent heart attacks before they happened. The thinking was: employ the solution before the problem occurs to prevent it from occurring in the first place. However, in 2020, two studies from the National Institutes of Health revealed that the use of these procedures is no more effective in preventing heart attacks than the combined efforts of medications and lifestyle changes, although stents can potentially be used to relieve symptoms of pain, if the patient is experiencing them. These days, cardiologists generally don't perform a cardiac intervention as a preventive measure before it is medically necessary. Of course, what is deemed "medically necessary" is a cause of much debate.

I saw a new patient recently who is thirty-five and has already had six stents placed. For someone so young to have arteries in such bad shape, there must be a genetic factor at play, and yet none of his doctors had checked. He'd had treadmill tests, but nobody had tested his physiological reactions to mental stress. He's never been on any common heart medications to try to prevent atherosclerosis, such as statins, PCSK9-inhibitors, ezetimibe, etc. Beyond those six stents, he's still got 59,999 miles of blood vessels that haven't been covered by a stent, but where plaque is continuing to grow nonetheless. Presumably, if he hadn't come to the Boone Heart Institute, he would have kept receiving stents until a heart attack finally killed him, long before his time should have been up. It's not feasible or ethical to continue employing such reactive, stopgap measures. Thankfully, after a thorough assessment through the Boone Heart Protocol, which allowed us to design an individualized treatment plan for him, we'll be able to stop the progress of his disease and avoid any further stent placements.

The only good stent is one that is placed during a true emergency, and then, only as a temporary solution while the patient begins aggressive treatments of the plaque with medication. At the Boone Heart Institute, we don't recommend the placement of stents, opting instead to use a combination of medications and lifestyle changes to reverse the progression of the disease. As we'll discuss further in Chapter 9, medications such as Repatha® have been found to be much more effective in healing not just the site of the volcano, but all of the plaque throughout the cardiovascular system.

Coronary Artery Bypass Grafting (CABG)

Coronary artery bypass grafting surgery (CABG) is the most advanced and drastic effort we can make to save a heart from the effects of ischemic heart disease. Often, when someone like Jim Fixx has passed away due to a cardiac arrest, they could have been saved by CABG—if only the problem had been discovered in time. In this highly successful procedure, cardiac surgeons take healthy blood vessels from other parts of the body, usually the legs, and create new pathways as needed around the blockage to restore blood flow. Crucially, this doesn't prevent the plaque from continuing to grow at the site of the original volcano of soft plaque, but simply reroutes blood around it. It doesn't prevent plaque from beginning to grow inside the newly placed blood vessels, either. Imagine if a bridge had collapsed, but instead of fixing it and improving the materials, a new one was simply erected somewhere else out of the same faulty construction materials. Everyone would think the builders were crazy. And yet, this is what we do every day with CABG.

At the Boone Heart Institute, we've only had two patients who needed emergency CABG, and it was because they had blockages of over 90 percent when they first came to see us. Now, I have many patients in their nineties who had heart attacks in their sixties and most of us have just about forgotten that they happened.

Angioplasty, stent placements, and CABG are the most common treatments for atherosclerosis, and they are the primary procedures we're trying to put an end to at the Boone Heart Institute. We should see the ongoing need for these procedures as a massive failure of the system, despite the brilliance of their invention. First, because it means that the disease went unnoticed for years, if not decades. Second, because in most cases, there are more effective ways to treat the disease through a combination of medications and lifestyle changes. And third, because these temporary solutions do nothing to halt or reverse the progress of the disease.

Valve Repair/Replacement

Beyond those surgeries for atherosclerosis, though, how do we fix damage to the heart muscle itself? The most common surgery is a valve repair or, if repair isn't possible, valve replacement. Dr. Myles Guber, my friend, patient, and the doctor who performed my mother's bypass surgery, specializes in repairing and replacing heart valves. It's an intricate procedure that may involve reattaching the delicate chordae tendinae, the network of spindly tendons that keeps the valve in place, or trimming or reshaping the existing valve to remove damaged or diseased sections. Cardiac surgeons prefer to keep as much of the original valve as possible intact, as it may lead to better long-term outcomes for the patient.

If, however, valve repair isn't possible and it must be replaced, the cardiac surgeon will use either a mechanical or animal (usually a pig's) heart valve instead. Mechanical valves may be more durable and the better choice for younger patients, but come with an increased risk of blood clots and may require the patient to be on blood thinners. For older patients, a pig's valve may be preferred because of the lowered risk for clotting and infection. Pigs' valves are remarkably similar to humans' and often make an ideal choice, although, like human valves, pig valves also wear down over time. Many such tissue valves can last up to twenty years, and we hope even longer with Boone Heart preventive techniques.

Valve repair or replacement surgeries are extensive, high-risk surgeries, as they involve the opening of the chest cavity so that a surgeon can access the heart to perform this technical operation. Understandably, it comes with long recovery times involving medications and therapy.

Ablations and Pacemakers

For those with AFib or PVCs that have been resistant to treatment with medications, an ablation may be a good option. In this procedure, a catheter is guided through the blood vessels to the heart, where electrophysiologists may use either extreme heat or extreme cold to purposefully induce scarring along the electrical route in the heart to disrupt the arrhythmias and reroute the electrical signal to its proper pathways from the SA node to the AV node.

For other patients, a pacemaker or implantable device may be inserted into the heart to help regulate its rhythm.

Heart Transplant

In extreme instances of end-stage heart failure, where other treatment options and surgeries have not been successful, an otherwise relatively healthy patient may be a good candidate for a heart transplant. In this case, the patient is placed on a waiting list in the hopes that a suitable match will become available. Often, patients wait months or years, depending on the urgency of their situation and the compatibility of prospective donor hearts. When a donor heart is identified, the surgery takes place as quickly as possible. After the failing heart is carefully detached from the arteries and blood vessels and removed, the donor heart is inserted and reattached. It is a painstaking process that takes several hours to complete. As you can imagine, the recovery is extensive, with patients remaining in the hospital recovery unit for weeks, and then months of medications and follow-up appointments to ensure that the patient is recovering well and that their immune system isn't rejecting the donation.

All the techniques, surgeries, and procedures described here are impressive, and the technology and skill of the cardiology teams that perform them cannot be celebrated enough. They are extraordinarily skilled in their efforts, and they become more effective every year. There is plenty of well-deserved recognition and reward for these cardiologists. You may well be grateful to know that these highly successful procedures exist and that you have access to them should the need ever arise.

Still, an ounce of prevention is worth a pound of cure. While every patient who undergoes these surgeries is undoubtedly grateful for them, I can assure you from speaking with hundreds of patients over the years that they would have preferred to never need them in the first place. If those patients could go back in time and identify the genetic factors that put them at risk, change their lifestyle habits, and take the right medications to prevent atherosclerosis or other forms of heart disease, they would. Moreover, the success of these procedures is largely dependent on the skill of the individual cardiologist or cardiac surgeon. Despite extensive training and practice, not every physician is equally adept and gifted. And even the best doctors have bad days. Even Peyton Manning and Tom Brady have lost Super Bowl games.

The problem is that prevention isn't as sexy, or as profitable. Although we have the right technology and imaging tests available to identify signs of heart disease sooner, these tests are not standard procedure. Too often, heart disease builds undetected throughout patients' blood vessels. And remember, *most* heart attacks occur in patients with less than 50 percent blockages in an artery with no chest pain, long before a stent would be deemed necessary. Although cardiologists' lifesaving interventions are incredible, the much greater obstacle to ending heart disease is the lack of early diagnosis and intervention.

You Can't Fix the Disease You Don't Know About

Lifesaving interventions for heart disease are only possible when we know about them—and most of the time, we don't.

My mother, Joyce, provides a good cautionary tale about the dangers of intervention over prevention. Joyce had diabetes and, as a result, was high-risk for heart disease. However, for years, her doctors only measured her blood pressure, blood sugar, and cholesterol—still the most common and woefully incomplete tests that most doctors use to assess the risk of heart disease. She was on a few medications for diabetes, but otherwise she was completely symptomless. That is, until the day she had a massive heart attack which revealed multiple blockages in her coronary arteries. She needed an emergency quintuple bypass.

That night, however, after her surgery, one of the bypasses closed back up. The cardiologist on staff decided that an angioplasty would be the best course of action to open it again. He threaded the catheter through a vein in her groin, but as it went through, it knocked off chunks of soft plaque that had built up in the rest of her blood vessels. One small piece of plaque traveled up to her brain and triggered a massive stroke, which left her with a permanent limp.

Thankfully, this story has a happy ending; under my care she lived another thirty healthy years. Not all patients are so lucky. But what I'm left with is the realization that her heart disease had been building for years, right under her doctors' noses. Instead of doing the work to discover this fact, they waited until something went very wrong. And even then, the treatment nearly killed her.

So why didn't her doctors know about her disease—and why don't they know about *most* heart disease? Unfortunately, the lack of early diagnosis of heart disease is almost inevitable, thanks to a confluence of systemic issues.

You Don't Know About What You Don't Test For

The lack of adequate routine testing, particularly imaging, is the main reason heart disease goes undetected for so long. It doesn't make sense, especially not when we are so watchful for other diseases. Consider that mammograms to check for breast cancer are standard practice for women over age forty. The American Cancer Society estimates the risk of getting breast cancer at about 13 percent for most women, and the mortality rate at 2.5 percent. Similarly, colonoscopies to screen for colorectal cancer are standard practice for both men and women above age forty-five, where the risk is about 4 percent, and the mortality rate is about .01 percent.

In contrast, cardiovascular disease is the number-one killer in the Western world. It's estimated that over 48 percent of people above age twenty in the U.S. *already have* heart disease, although most of them don't know it yet. The American Heart Association reported in early 2024 that "someone dies of CVD every 34 seconds in the U.S."

We can and must do better.

It is simply not enough to test blood pressure, cholesterol, and blood sugar for the simple reason that they cannot give us an accurate indication of the amount of plaque building inside someone's arteries. And we cannot wait until someone begins experiencing symptoms to employ CT angiograms or ultrasounds of the arteries, because by that point, the disease has already taken hold. If we want to end atherosclerosis, everyone should have an initial screening with a quick, plaque-detecting ultrasound of the carotid arteries at age twenty-five. If you know of relatives who passed early in life from heart disease, then getting a genetic panel in your twenties is a good first step, and a one-time cost that will help you live a longer and healthier life. By age thirty-five, everyone should have a full workup to screen for heart disease as the standard of care. We'll talk more about this in Chapter 9.

There are other contributing factors, too, that make comprehensive testing for heart disease more difficult. Healthcare is fragmented, with many individual specialists operating in silos. Often, as patients are referred from one specialist to the next on the hunt for a diagnosis, important information gets missed or dropped off. The waitlist for specialists is often months, meaning valuable time may be wasted as patients wait for an initial consultation with a new provider, then even longer to schedule appointments for labs and tests. Often, the only point of consistency in a patient's journey to diagnosis is the patient himself or herself. And most patients don't know what to ask for.

Patients are left wondering what step is right to take next, with the added burden of not knowing what information to communicate to the next person, instead of having access to a qualified and knowledgeable healthcare professional who can help guide them through every step of the process. You may go to your general practitioner (GP) for shortness of breath, only to be referred to a psychiatrist for anxiety, or an allergist to check for certain allergies. Heart disease is a lookalike problem; often, the symptoms of ischemic heart disease look much like the symptoms for other diseases. It's also closely interrelated to other conditions like liver and kidney disease, diabetes (as seen with my mother, Joyce), obesity, and even cancer and Alzheimer's. It's easy for patients to be led down a path away from the right diagnosis, especially in a system where specialists operate in silos, without consistency or consolidation of healthcare records and other vital information regarding the patient.

This wandering path to diagnosis is further hampered by the tendency to place blame (and prevention) for heart disease on lifestyle.

Blaming Lifestyle

Across the medical community and amongst the general public, there is a disproportionate amount of responsibility placed on patients' lifestyle choices as contributors toward heart disease, and therefore, a disproportionate emphasis on lifestyle changes as the solution. While it's true that lifestyle factors, such as smoking, alcohol use, lack of exercise or excessive exercise, and dietary choices may be risk factors for heart disease, one of the primary goals of this book has been to expose the arguably more pressing risk factors that too often get ignored, mainly genetics. Even many doctors don't understand the pivotal role of genetics in the origins of heart disease and leave it up to their patients to try to prevent a disease they may be predisposed to getting despite making good lifestyle choices. As you already know, for those with LP(a) or any number of genetic predispositions, eating healthy and exercising moderately on a regular basis won't be enough to prevent heart disease.

Assigning the responsibility for heart disease to lifestyle choices is a convenient shift in responsibility, and allows many providers to avoid looking deeper for early signs of disease and treatments. It's the easiest possible solution for everybody. Because there's still a lack of general understanding around the complex risk factors for heart disease, many doctors don't know where to start other than to refer patients to a cardiologist. And as already seen, the typical cardiology appointment often misses the signs that are hiding in plain sight. Even if a sign of possible heart disease is uncovered, the right next steps to take to prevent its progress toward an acute cardiovascular event are often uncertain. It's almost easier to wait until an acute event (like a heart attack) occurs; then, at least, the action steps will be clearer.

Prevention is far more complex than repair. No matter what caused the problem, the solution is almost always the same: clear the pipes, replace the leaky valve, or fix the wiring. Raising awareness of the right testing and diagnostic measures for heart disease, and providing physicians with concrete steps to take to prevent its progress, is one of the main reasons we developed the Boone Heart Protocol, which we'll talk about in Chapter 8.

Lack of Access and Discrimination

Yet another contributing factor to heart disease's prevalence is that many patients cannot seek treatment or start the laborious process to discover the root cause of their symptoms, due to circumstances or stigma. For too many patients, lack of transportation, work schedule conflicts, lack of health insurance, financial stress, and language barriers limit access to healthcare and prevent them from seeking timely treatments.

In addition, many people in rural and urban communities alike live in "healthcare deserts." Healthcare deserts are defined in a number of ways, but generally are characterized by a pervasive lack of access to primary care providers, adequate staffing of healthcare workers, timely appointments, hospitals and treatment centers (and/or hospital beds), pharmacies, and low-cost healthcare clinics. In a 2021 report on healthcare deserts, the research team at GoodRx found that over 80 percent of counties in the United States meet at least one of the characteristics for healthcare deserts. Citizenship status also plays a significant role, as noncitizens tend to have limited access to healthcare.

And for people in many of these areas, the risk of heart disease is even higher if they also live in a food desert with insufficient access to healthy foods. A 2019 study found that living in a food desert and having low income led to poorer health outcomes for those with CVD.

Further, because of the history of discrimination and bias against minorities in the medical community, there is often an understandable lack of trust in healthcare providers amongst people of color.

Lack of trust in the medical community is a pervasive challenge, and it's only gotten worse since the COVID-19 pandemic. In 2021, a study initiated by the American Board of Internal Medicine (ABIM) and conducted by NORC at the University of Chicago, found a discrepancy between how much physicians versus patients trust healthcare professionals (and the healthcare system as a whole). In general, physicians trust other doctors and nurses far more than patients do, with 80 to 90 percent of doctors reporting that they "completely trust" or "somewhat trust" the doctors and nurses on their teams, compared to only 68 to 78 percent of patients.

The study also found that lack of insurance coverage and level of income were among the most prevalent reasons that patients experienced discrimination in the healthcare system, besides race/ethnicity and gender. And the last reason why the healthcare system often doesn't discover heart disease until it's too late is because it is paid not to.

Follow the Money

Upton Sinclair famously wrote, "It is difficult to get a man to understand something when his salary depends upon his not understanding it." The same truth applies in healthcare: As long as there is more money to be made by prioritizing the Hail Mary last-resort measures to slow heart disease rather than preventing it, our healthcare system will continue to do so.

Think about it. No one gets paid if you avoid processed foods, exercise regularly and moderately, live a long, healthy life, and die peacefully of old age in your nineties or beyond. Preventive measures like gym memberships, nutritional coaching, and testing are treated as elective, "nice-to-haves" rather than "must-haves." The best cardiovascular patient for hospitals and medical providers is one who has a sudden heart attack or stroke, rides in an ambulance to the hospital, receives a lifesaving bypass surgery, and spends the rest of his or her life in a nursing home on expensive medications having procedures done periodically. Healthcare providers profit far more from treating patients with severe or advanced disease than they do from preventing those diseases in the first place.

The system isn't interested in your health; they're only interested in the cost of your health. The money isn't in preventing disease; it's in *fixing* disease.

To illustrate the financial stakes and possible gains for healthcare, let's look at just the costs of coronary artery bypass grafting (CABG). The costs for any cardiac procedure vary greatly across the country, but the 2022 International Health Cost Comparison Report from the International Federation of Health Plans found that the median price (with insurance) for CABG in the United States in 2019 was $76,384. Meanwhile, if you don't have healthcare coverage, the costs of the same procedure are even more prohibitive. A 2016 study of "out-of-pocket" costs for CABG in the U.S. found the mean price to be $151,271. Keep in mind, too, that these costs do not include the costs of physical therapy, follow-up procedures and appointments, and the expensive regimen of medications that most patients will require after receiving these treatments.

According to the non-profit Kaiser Family Foundation (KFF0, Americans owe over $220 billion in medical debt. And considering that heart disease is the number-one cause of death in the U.S., it's safe to assume that a good chunk of that debt is held by those who are suffering from CVD. In fact, in 2019, the American College of Cardiology shared in a press release that over 45 percent of "adult atherosclerotic cardiovascular disease (ASCVD) patients suffer financial hardship related to their medical bills, including many who cannot pay their medical bills at all."

In comparison, the costs of genetic testing to check for LP(a) and other markers of heart disease are minimal, typically in the range of

$100-$300 depending on whether insurance covers it. Even if insurance doesn't cover it, this testing is well worth the cost, since it's just a one-time test. Get the information early, and you can take the steps you need to prevent heart disease for a lifetime. Likewise, the cost of imaging is a few hundred dollars, and the average cost of medications ranges from an average of $20/month for statins and up to $560/month for PCSK9 inhibitors, like Repatha® (which you'll learn about in Chapter 9). At the Boone Heart Institute, we are working hard to drastically reduce the costs of these lifesaving preventive measures and make them more affordable and accessible across the country. To some, these prices may seem prohibitive, but in comparison to the cost of treatment, they are minimal. Frankly, we can no longer afford *not* to prioritize the prevention of heart disease over the treatment of it.

And the ability to enjoy a long and healthy life, free from heart disease?

That's priceless.

The Next Frontier in Heart Disease Prevention

We need a systems-wide realignment towards prevention. Imagine if firefighters were trained to reconstruct houses instead of save them, and as a result we waited to put out fires until entire rooms or the entire building collapsed. That would be absurd. Why, then, do we wait for our health to be permanently compromised—or worse—before we're willing to take action to diagnose and address the underlying issue? Fire departments don't just wait around for the alarm to start ringing. We need to invest in fire prevention: better alarms, sprinkler systems, electrical upgrades, and fire-safe walls.

While it's beyond the scope of this book to change the U.S. healthcare industry, at the Boone Heart Institute, we believe a cure for atherosclerotic heart disease is closer than ever before. It's feasible that within our lifetime, we'll see the life-threatening power of heart disease dwindle in the way that so many past killers have been rendered impotent, including smallpox, tuberculosis, and even HIV/AIDS. In Part II, we'll explore the exciting new scientific advancements that pose the greatest opportunities for curing every person of atherosclerosis and eradicating it altogether.

PART II

How to Eradicate Heart Disease

CHAPTER 8

How Diseases Get Cured

On July 1, 1983, a young man was admitted to Good Samaritan Hospital in Portland, Oregon, presenting with a list of symptoms that sent a shiver down every doctor's back: Fever. Weight loss. Pneumocystis pneumonia. Kaposi sarcoma, a cancer of the lungs. Sweats and chills.

AIDS had reached us.

Although the mysterious disease had been spreading slowly across the United States throughout the seventies, and had reached a critical frenzy in major cosmopolitan areas like New York City and San Francisco by the early 1980s, this was the first confirmed case in Oregon. Just two months earlier, virologists at the Institute Pasteur in Paris had finally isolated the cause of this illness as the Human Immunodeficiency Virus (HIV). So much was still unknown about it, including how it was spread. Some doctors refused to see the patient altogether. Others entered his room dressed like the Tin Man from *The Wizard of Oz*. There was talk of having to burn down the hospital.

I was the intern doctor assigned to watch over him.

For thirty-one days, he received the same treatments and explorations that doctors in bigger cities had already tried (which had failed): IVs, sputum analysis, stool analysis, X-rays, steroids, statins. He was examined by thirty different specialists, who came to see him at all hours of the day and night, ordering new tests and trying to find the clue that would lead to a cure. I worked 138 hours a week, round the clock, acting as the liaison and primary point of contact for his care at the hospital. By the time he arrived at the ICU where I was working, he had already been intubated in the emergency room; otherwise, if his breathing reached a critical point, I would have had to give him mouth-to-mouth. As it was, all we could do was treat his symptoms, make him comfortable, and wait for the end we already knew was coming. (In many ways, this is not unlike the standard medical care for patients with atherosclerotic heart disease.)

When a disease has no cure, that is all you can do: treat the symptoms and hope for the best. We worked furiously to address the peripheral effects of this mysterious virus, but we could not see or treat the disease itself. We were overcome with fear and helplessness.

I remember in those weeks, we had innumerable meetings with his family. They knew him as a strong, handsome, vigorous young man, full of life and energy. I remember the shock on their faces as we told them: *Your son is going to die. There's no cure.* They hadn't even known he was gay.

On July 31, after a short and bitter battle, he passed away.

It was one of the most stressful and hectic times of my professional career, and yet the memory of that man and his family fighting for his life left an indelible mark in my mind, impressing upon me the urgency of fighting for a cure, no matter the obstacles.

Over the course of the following decades, I watched public perceptions of AIDS change drastically. I remember on April 9, 1987, when Princess Diana made history with a handshake. At a time when most AIDS patients wouldn't volunteer to be photographed shaking the princess's hand, for fear of being recognized and shunned, one brave man (his back turned to the camera) agreed to help change the tide of the public stigma. Later, she visited AIDS hospice patients all over the world, giving hugs as often as she could. Thanks to her efforts, and the efforts of many brave doctors and researchers, misinformation about this terrible disease was finally put to rest and the stigma around HIV/AIDS has now lessened significantly.

Not only that, but I watched as the world united in the quest to find a cure. The research and advocacy conducted by the National Institute of Allergy and Infectious Diseases (NIAID), led by Dr. Anthony Fauci, secured the widespread availability of new drugs for AIDS patients, first AZT and now antiretroviral therapy (ART). In the early eighties, HIV/AIDS was a death sentence, with most patients dying within a year of diagnosis. Now, however, HIV patients receiving ART can expect to live long and relatively healthy lives. Thanks to stem cell transplantation, several have even been cured, illuminating further hope not only for other patients living with HIV/AIDS, but for the curing of all diseases.

Consider this: At the height of the AIDS epidemic in 2004, two million people worldwide died from the virus and subsequent infections. Today, heart disease is killing nearly eighteen million people across the globe each year—*nine times* as many lives as AIDS at its peak. And yet, we don't hear breaking news every day about the climbing death tolls from heart disease. The media doesn't report when a major breakthrough in technology or drug research or preventive medicine occurs. Today, the drugs most successful in treating heart disease are better known for their appetite-quenching side effects and the resultant weight loss they bring about. The urgency that propelled global task forces to unite across cultural and language divides to find a cure for AIDS simply isn't there with heart disease.

And yet, I believe we're closer now to a permanent solution to heart disease than we ever have been, and with the new technologies and scientific advancements that are happening (although most people don't hear about them), there's no reason that heart disease shouldn't quickly go the way of smallpox, AIDS, tuberculosis, polio, and hundreds of other diseases that were once death sentences. In fact, at the Boone Heart Institute, we're already seeing that many patients with advanced atherosclerosis are recovering from the disease and going on to live healthy lives for decades more.

It's time to find the cure.

Is a Cure for Heart Disease Possible?

In the pursuit of a cure for any disease, the goal is always total eradication—that is, for the disease to be globally and permanently wiped out so that no one will ever need intervention or treatment for it again. To date, the World Health Organization has declared only one human disease to be officially eradicated: smallpox. With many other, mostly communicable (i.e. infectious) diseases, we've gotten very close with the development of vaccines, intensive drug therapies, and widespread education and prevention.

For millennia, infectious diseases posed the greatest threat to humanity. Now, as Western society has become better equipped to eliminate the threat of infectious disease, noncommunicable diseases have risen to the top of the list of killers, including heart disease.

The journey to curing noncommunicable diseases appears to be a longer one, though, as these diseases seem to arise "spontaneously" from inflammation and oxidation within the body. In fact, many doctors believe that noncommunicable diseases are not curable at all; the best we can hope for is their continual management. In a 2020 paper sounding the alarm on noncommunicable diseases (NCDs) and urging people to rally to pursue treatments and cures, authors Florencia Luna and Valerie Luyckx warn against "therapeutic nihilism," or the hopelessness that some doctors and researchers may feel in fighting for a cure for NCDs. They write, "Some NCDs, such as some forms of cancer, may also be curable. In contrast, some CDs, such as HIV, are not curable, or others such as tuberculosis may require long courses of treatment. These circumstances have not prevented treatment efforts for these infections." The biggest threat to finding the cure for noncommunicable diseases, starting with atherosclerosis, is the mistaken belief that *it can't be done*.

I want to be completely clear here. Atherosclerosis can be *eradicated* from humanity as a premature cause of death. Yes, we have the lifestyle interventions, surgeries, drugs, and education to not only prevent it but also to reverse it once it's started by treating all sixty thousand miles of blood vessels. But eradication goes further than that, eliminating the disease completely as a threat to human life, with breakthroughs on the horizon like vaccines and gene editing therapies. The advances we're making in the eradication of atherosclerotic heart disease may chart the course towards a cure for other noncommunicable diseases as well, like cancer, dementia, Alzheimer's, autoimmune diseases, kidney and liver disease, and diabetes, among others, as we'll talk about in Chapter 10.

It's worth noting that the other forms of heart disease, particularly problems with the muscles and valves and the electricity, will still persist, especially as atherosclerosis is eliminated. Our hearts will be expected to work longer than they ever have, leading to the eventual wear-down of the muscles and valves, and the occasional malfunction of the electricity. For those problems, we continue to have effective surgical treatments that will remain available and even improve as those surgeries become more common. But atherosclerosis, the Trojan horse of heart disease and the leading cause of death worldwide, can and will be eradicated very soon: within your lifetime, if not in mine.

Luna and Luyckx argue convincingly that there is no moral grounding for the persisting myth that noncommunicable diseases can't be cured. Both CDs and NCDs may have acute and chronic symptoms; both have equally effective and cost-efficient methods of treatment available; both may be caused by environmental factors or individual choices; and, with new understanding of the role of genetics and social determinants or risk factors (like lack of education, or living in food or healthcare deserts), we now know that both may be passed from one person to the next, making NCDs just another type of "infectious" disease.

It's well past time that we started treating NCDs, starting with atherosclerosis, with the same urgency and single-minded attention that we gave to the Spanish flu, the COVID-19 pandemic, and other epidemics. In fact, we can learn a lot about how to cure heart disease and chart a path for success by looking at the recent history of our search to cure communicable diseases, like tuberculosis.

How to Cure a Disease: the Story of Tuberculosis

Tuberculosis was a leading cause of death worldwide for nearly a thousand years, although advancements in treatment and healing happened quickly, beginning in the late nineteenth and continuing throughout the twentieth century. The search for a cure happened (and still continues) in a series of phases that aligns well with our search for the cure for heart disease, starting with the first encounters.

Phase 1: What Is It?

Tuberculosis has been around for so long that different cultures at different times have had different names for it, including *phthisis* (in ancient Greece), *scrofula* (in Europe during the Middle Ages), "the white plague" (in the eighteenth century), and *consumption* (in Europe and the Americas during the nineteenth and twentieth centuries).

Doctors were sparse in these times, and those that did exist were limited in their knowledge and research of the disease. Most people, especially the poor, lived in overcrowded boarding houses and worked side-by-side in factories with unhygienic living and working conditions, allowing the rapid spread of the disease from one person to the next, although for most of history they did not realize tuberculosis was contagious. To cure scrofula, hundreds would line up to receive the "royal touch" from the kings and queens of Europe, as it was thought that they were imbued with special healing powers as part of their divine authority.

The only other known cures during this phase of the disease were home remedies, including mint and green tea, pineapple, milk, and fresh air, or dubious concoctions of snake oil, including turpentine, cod liver oil, vinegar, and hemlock.

During the age of consumption especially, Western society held a very romantic or idealized notion of the disease, as it attacked the young poets, musicians, and artists whose work defined the culture. Throughout nineteenth-century Western literature, you can find glowing descriptions of delicate, frail, and almost angelic consumptive patients, to the point that the poet Lord Byron wrote that he wished to die of consumption, and women even ate sparingly in order to achieve the "consumptive look."

Not much was known about the disease until the nineteenth century, when technology and scientific experimentation finally began to shed new light on the epidemiology of it.

Phase 2: Early Attempts at Treatment

The first real step towards understanding tuberculosis was taken by French doctor, Rene Laënnec, who invented the stethoscope to aid him in his chest examinations of tuberculosis patients. Through decades of listening, he described the particular sounds of fluid and pus in tuberculosis patients' lungs, which helped later doctors begin to diagnose the disease more quickly. This was how he diagnosed his own case of consumption, which unfortunately killed him in 1826 at age forty-five.

Then, in 1882, German physician and microbiologist, Hermann Heinrich Robert Koch, demonstrated that he had identified the cause of the disease as *Mycobacterium tuberculosis* (Mtb), which is spread through infected particles or droplets from one person to another. In that presentation, he put forth four criteria that are still used to this day as the gold standard for determining the etiology, or cause, of a bacterial disease. Now known as "Koch's Postulates," the four criteria are:

> "(1) The microorganism must be found in diseased but not healthy individuals; (2) The microorganism must be cultured

> from the diseased individual; (3) Inoculation of a healthy individual with the cultured microorganism must recapitulate the disease; and finally (4) The microorganism must be re-isolated from the inoculated, diseased individual and matched to the original microorganism."

Koch showed that tuberculosis is a slow-moving disease, with the bacterium reproducing very slowly, only once every twenty-four hours. It can take weeks or months before an infected person begins to show signs of the disease.

Now we know more about the biological process occurring as the disease progresses. Once the Mtb reaches the lungs and begins to settle in, the body's warrior white blood cells, macrophages, come in to devour and destroy the errant bacteria. The macrophages initiate an inflammatory response to kill Mtb through heat, but it doesn't work. Instead, as more and more immune cells flock to the infected site, lesions begin to form. Sound familiar? This is nearly the same process as atherosclerosis, described in Chapter 2.

This was a major breakthrough, as it finally pointed the way toward ending the epidemic through social isolation of those afflicted with tuberculosis. In the mid-1800s, sanitoriums arose as places for wealthy tubercular patients to rest and avoid exposing their families and others to the bacteria.

The theory that the quality of air made a difference in treatment was already popular by this time. Since tuberculosis is a disease primarily of the lungs, people believed that by breathing fresh air, especially at high elevations, you would be healed. Tuberculosis patients often traveled to higher-altitude destinations to experience the dry, healing air, and now sanitoriums were erected in the same places. Many of the hospitals around Denver were first founded as sanitoriums, particularly for Jewish patients, leading to Denver's reputation as "the World's Sanitorium."

Some people did heal at the sanitorium, if only because the rest supported their immune system in fighting off the infection and inflammation. But just as often, patients didn't recover.

Beds at sanitoriums were in high demand, but often long periods away from family weren't feasible for tubercular patients, so the other treatment option that arose during this period was surgery. The thinking was that rest encouraged the healing of the lungs, but if the lungs weren't showing signs of improvement, or the patient didn't have time for a prolonged period of rest, then there must be another way to enforce rest. Therefore, the late 1800s and early 1900s saw the rise of pulmonary collapse therapy, or *therapeutic pneumothorax*, where one lung was purposefully injected with nitrogen to force it to collapse, allowing the lesions in the lungs to heal. This forced collapse also stops the spread of the bacteria. The injections had to be repeated every two to three weeks, for periods of two years or more. This surgical intervention was and is highly successful, and it is still used today as a treatment for drug-resistant TB. Other surgical interventions included resections, where the diseased portion of the lung was removed, and lobectomies, where in some cases, the entire lung was removed.

Lifestyle interventions and surgeries are usually the first place to start with fighting a disease, and this is what we've seen with atherosclerosis in the last hundred years. However, once lifestyle changes have been shown to, at best, mitigate the effects of the disease, and surgeries have proved too invasive, costly, or inaccessible, then drugs and vaccines can aid in the search for a more permanent solution.

Phase 3: Drugs and Vaccines

After his famous discovery of the *Mycobacterium tuberculosis*, Koch continued his experiments in the hopes of finding a cure. In 1890, he developed a substance he called tuberculin, which he thought could be injected into positive tubercular patients to halt the progress of the disease. Unfortunately, it proved to be ineffective there, although in 1908, pediatrician Clemens Von Pirquet discovered that tuberculin could be used to identify children with latent tuberculosis when applied to the skin.

Then, in 1921, at the Pasteur Institute in Lille, France, Albert Calmette and Camille Guérin discovered the first vaccine for tuberculosis, which they created from *Mycobacterium bovis*, a form of tuberculosis found in cattle. Their vaccine, called BCG (Bacille Calmette-Guérin), became the first and is still the only vaccine used to inoculate against tuberculosis. In his article, "The History of Tuberculosis," Thomas Daniel writes of this discovery: "An infant born of a mother dying of pulmonary tuberculosis and placed in the care of a tuberculous grandmother was the first recipient. The child survived and did not develop tuberculosis. Over the next seven years, more than a hundred thousand children were immunized, including Calmette's children."

The next major discovery came in 1944 at Rutgers University, when Dr. Selman Waksman and his research student, Albert Schatz, isolated one of the first antibiotics found to not be toxic to humans, which they named streptomycin. Researchers at the Mayo Clinic tested the new antibiotic on guinea pigs infected with tuberculosis, finding that they responded quickly and that their conditions improved rapidly with the drug. Dr. Waksman was asked to develop more streptomycin so that it could be tested on humans, again with positive results. Within ten years, the FDA had approved streptomycin as a cure for TB, and a factory was developed for mass-producing the drug. In 1952, Dr. Waksman received the Nobel Prize for his efforts in this remarkable advancement.

However, in the decades following, it became clear that some forms of TB developed a resistance to streptomycin. As a result, new drugs had to be developed, and today a combination of drugs is the most common treatment for TB. The CDC says that "Today, four drugs are used to treat TB disease: isoniazid (1951), pyrazinamide (1952), ethambutol

(1961), and rifampin (1966)." In cases where patients fail to respond to drug treatments, surgeries such as therapeutic pneumothorax and pulmonary resections are still used.

Drug-resistant TB continues to be a challenge in many areas of the world. In fact, the World Health Organization (WHO) reports that tuberculosis remains among the top ten causes of death in low-income countries.

In first-world countries, the presence of TB has been dramatically reduced, to the point of being nearly eradicated. The BCG vaccine is no longer recommended or needed in the United States, although it is delivered to infants and children in places where TB is still common. Western countries do remain vigilant in watching for signs of recurrence, however. As a doctor, I am required to get a TB test every year on the off chance that I somehow am exposed. If you're watching the commercials for new therapies for rheumatoid arthritis and eczema, such as HUMIRA® or SKYRIZI®, you may hear the narrator say that those treatments could reactivate latent tuberculosis, if you happen to have it. Anybody with those diseases has to get a TB test.

Like TB, atherosclerosis is in Phase 3 of curing. We have the drugs to halt its progress, but unlike TB, which develops from a live bacterium that can develop a resistance to the drugs, atherosclerosis will always respond to the drugs we give patients, like statins and PCSK9 inhibitors (Repatha, Praluent, Leqvio). There is also a new vaccine targeting PCSK9, the protein that regulates our LDL cholesterol, to improve the number and function of LDL receptors in the liver, in order to reduce the overall cholesterol burden in our bodies. The vaccine has already shown tremendous results in tests with monkeys and could be key in the future prevention of plaque development.

But a future is coming, and may already be here, in which we can change the course of diseases like tuberculosis and atherosclerosis, thanks to new stem cell and gene therapies that are being developed.

Phase 4: Future Cures Through Stem Cell and Gene Therapies

Gene editing through CRISPR ("clustered regularly interspaced short palindromic repeats") technologies represents an exciting new possibility for curing atherosclerosis and detecting latent tuberculosis. CRISPR is a technology that allows geneticists to modify or "edit" parts of the human genome by altering sections of our DNA. It's a controversial method, and yet it poses the greatest potential for a permanent cure for both noncommunicable and communicable diseases. Already, studies have been conducted showing that gene-editing through CRISPR significantly reduced the levels of LDL cholesterol and PCSK9 in patients with heterozygous familial hypercholesterolemia, a serious genetic mutation in which the body cannot produce or produces less of the LDL receptors necessary to process and dispose excess LDL cholesterol. The research around gene-editing is still nascent, but is growing rapidly, and someday soon, we may no longer need the PCSK9 inhibitors we now rely on. That, combined with education, atherosclerosis-reversing drugs, and lifestyle improvements, will finally start to decrease and eventually eradicate heart disease as a threat to human life, just as we've seen with smallpox and tuberculosis, which have been eliminated in the United States, if not quite yet eradicated worldwide.

It took a thousand years to stop tuberculosis. Thankfully, heart disease has had a far shorter time at the top of the list of global killers—a hundred years instead of a thousand. We've rapidly learned the etiology of the disease, and how to stop and even reverse its destruction. Soon, atherosclerosis will no longer be the death sentence it has been for the past century.

Of course, you can't cure the disease if you don't know you have it.

Eradication starts with one person at a time. At the Boone Heart Institute, we've seen hundreds of patients walk through our doors in advanced stages of atherosclerosis, and with the individualized knowledge and custom treatment plans we've created through the Boone Heart Protocol, they continue walking through our doors decades after they might have died. We have proof that anyone living with heart disease can be completely healed and go on to live an abundant life free from any signs of plaque.

In the next chapter, I'll tell you how.

CHAPTER 9

The Boone Heart Protocol

I arrive every morning at an inconspicuous building tucked away in the middle of a small medical complex in the heart of the Denver Tech Center neighborhood. The lobby is quiet and warm, with calm colors designed to welcome patients and lower their stress levels. On any given day, I may walk in to be greeted by a CEO from Dubai, the head of a major nonprofit foundation, a former ambassador, a famous sports legend, or someone I've never met—long-time patients and new ones—all driven here by a desire to take hold of their future and ensure that they'll continue to be around for a very long time.

Welcome to the Boone Heart Institute.

Through the lobby and into the main office, the hallways are lined with photographs and autographs of stars from every professional sport, actors and actresses, and even an astronaut or two—a tribute to many of the lives we've saved, and a reminder of each era in our history. I see former Denver Broncos players that we worked with through the Denver Broncos Cardiovascular Prevention Program. I think of every Hall of Fame player I met in the NFL Alumni program. I pass autographs of Hall of Famers from the Cardinals—Bob Gibson, Lou Brock, Red Schoendienst, Stan Musial—and think of my dad, who grew up in Kansas when baseball was the national pastime. He listened to every Cardinals game on the radio at my grandfather's gas station. In 1964, I even got to go to the World Series game with my father and grandfather. When we started testing the Rockies coaches, they didn't like seeing all the Cardinals autographs, and so they started sending us Rockies memorabilia signed by players: Charlie Blackmon, Carlos

Gonzales, and others. Representing the Denver Nuggets is an autograph from Nikola Jokić, whose brothers are also now patients of ours.

One of my favorite pictures in my office is of Muhammad Ali talking with Bill Russell, the famous basketball player for the Boston Celtics, who came and lectured at our college in 1970. I got to have dinner with him after the event, and Russell spoke about racism and social activism, reminding me that basketball was just his job; his real contribution was in fighting against racism in sports. Behind my desk you can also spot autographs from some of the greatest football players in history: Peyton Manning, Tim Tebow, Bart Starr—all players we tested at one time or another.

These pictures and signatures are constant reminders not only of the work we have done, but also of the work still left to do. Many of the incredible athletes pictured on my walls died far too early because we didn't have the lifesaving drugs or imaging technology available at the time to save them. Every day, their pictures remind me of the commitment we've made to ending atherosclerotic heart disease.

It's an immense privilege that our team has been able to work with so many world-class athletes, coaches, businessmen, and pioneers in every field. And yet, living free from atherosclerosis is not a luxury reserved for a small elite group of (mostly male) leaders in the United States. One of the goals of this book is to ensure that every person across the globe—men, women, and children— has both knowledge of and access to the affordable and high-quality interventions that cure people of heart disease. Many of our high-profile clients' lives have been changed, and their support and advocacy have enabled us to spread the word much further than we might have alone. We've worked with everyone from the Major Leagues to high school teams, always bringing cutting-edge technologies and research to support our efforts. The Boone Heart Institute is known for taking risks, for pioneering new advances, and being the first to adopt whatever is proven to work. Many of the methodologies and techniques we adopted ten or twenty years ago are now commonplace. Statins that used to cost patients hundreds of dollars a month are now available at a fraction of the cost. And what's exciting is that there is always something new on the horizon. Next year, the technology and knowledge will only be better than it is now. There will be new drugs on the market, new vaccines, new surgeries, and new advancements that will keep us on the cutting-edge of curing heart disease.

I'm proud to say you couldn't take the President of the United States to any place better than ours, and at the same time, the poorest person in Washington, D.C. could benefit from the exact same treatments.

In this chapter, I'll take you behind the scenes of what we do at the Boone Heart Institute and introduce you to the Boone Heart Protocol, our bespoke framework for finding atherosclerosis before it strikes, discovering the root causes, and aggressively reversing it.

The Boone Heart Protocol

Recently, a colleague of mine, Dr. Thomas Haffey, told me, "Everyone in the western world will die of atherosclerosis—every one of us. But for half of us, something else gets us first." Dr. Haffey is the president of the Denver chapter of the American College of Cardiology and the president of the Southwest chapter of the National Lipid Association. He sees firsthand every day the effects of atherosclerosis, and believes there is no such thing as a "sudden" heart attack, although one might catch you by surprise. As we've seen in previous chapters, there are always signs of plaque and an impending heart attack before one happens, but only if you know how to look for those signs. His words struck me as a powerful note about the important work we and others like him are doing to educate doctors and patients on heart disease and stop atherosclerosis in its tracks.

At the Boone Heart Institute, we developed a simple protocol for helping us diagnose, understand, and treat heart disease in every patient who walks through our doors. This protocol is framed around four questions that guide us to a personalized treatment plan designed to reverse any existing heart disease and prevent the formation of new atherosclerotic plaques throughout the blood vessels. A more business-minded person might think to protect this information, patent it, and share it only at a great cost. But that runs counter to our mission at Boone Heart. To truly stop heart disease, we need to share our formula with everyone.

The four questions are:

First, do you have the disease? Beyond testing the traditional indicators of heart disease, like blood pressure, cholesterol, and blood sugar, there are several ways we can detect whether a patient has plaque through precise and personalized imaging.

Second, why is it there? What combination of genetics, lifestyle factors, and other environmental or biological factors are coming into play? Understanding each person's unique biology in the context of their lives helps us prescribe the right treatment and ensure their health for as long as possible.

Third, we ask: Is the plaque vulnerable? That is, is the hot lava in your arteries due to erupt in a myocardial infarction? Inflammation is the key indicator here that a heart attack, stroke, or pulmonary embolism may be waiting in the wings for the right stressful moment to strike. Again, the results here will help dictate the specific course of treatment.

Finally, when we've determined the extent of the disease and the current threat of an attack, we consider a treatment plan: What do we do about it? The right course of action will be slightly different depending on the circumstances, but every person with heart disease will need to consider the right medications, lifestyle changes, and stress management techniques.

I want to take you step-by-step through these questions and considerations, so that you can advocate for your health and find the support you need to fight atherosclerosis, or prevent it altogether.

Do You Have the Disease?

When people have their annual physical or check-up, doctors typically take three important measurements to determine whether they might be *at risk* of heart disease: blood pressure, cholesterol, and blood sugar. If any one of these comes back in an abnormal range, the patient will likely receive a referral for a follow-up appointment with a cardiologist for further testing.

The problem with this approach is that usually, by the time you're suffering from hypertension (high blood pressure), or your LDL cholesterol shows in unhealthy ranges, or you're diagnosed as prediabetic, plaque has already been building in your blood vessels for years. The stage has already been set for atherosclerosis to thrive. There is an overreliance on cholesterol tests, particularly. As authors from the Harvard Medical School write, "Research has shown that only about 50 percent of the people who have heart attacks have high LDL. If LDL levels are supposed to be an alarm, then it's not going off for half of those who might benefit from a wake-up call."

Why let it get that far? Especially when we have all the testing available to catch plaque much earlier and confidently diagnose the disease without having to wait for follow-up appointments and tests? High cholesterol isn't the disease. High blood pressure isn't the disease. Diabetes is not the disease. We don't have to test these potential *indicators* of heart disease, which may be signs of something else as well, when we could definitively rule out or diagnose the presence of atherosclerosis itself through imaging.

Thanks to Rene Laënnec and his invention of the stethoscope, for decades that was the best way for doctors to get insight into what was going on beneath the surface of the skin. Most cardiologists were trained to listen to people's hearts for the soft *whooshing* sound that might indicate narrowed blood vessels where plaque may be building up, particularly the sound of bruits in the carotid arteries. But I personally haven't used my stethoscope in twenty years because we have much more sophisticated imaging available to us. It's almost criminal not to take a look inside, now that the technology is so widely accessible.

The two root causes of atherosclerosis (and all disease, as we saw with tuberculosis as well) are inflammation and oxidation. This is essentially what we're looking for when we go searching for heart disease. The first step is an ultrasound of the arteries in the neck, in which we look for carotid plaque and measure the inner lining of the artery in a carotid intima-media thickness (CIMT) test. The CIMT measurement allows us to detect the earliest stages of the disease and measure its reversal with repeat scans. These tests are easy to do, take very little time, and are low cost. The presence or absence of plaque on this initial scan is usually our first discovery of atherosclerosis. Most doctors' offices, and all hospitals in the United States, already have these ultrasound capabilities, and we could be using them more to diagnose heart disease early on. However, they are usually only incorporated when you have symptoms, or advanced surgical disease.

Ideally, starting at age twenty-five, every person would have an ultrasound of their neck and legs (often an early breeding ground for plaque) as part of their routine annual physical. Studies have shown that atherosclerosis may start, without symptoms, in a patient's twenties or thirties. I find that by the time most people reach thirty-five to forty and often have kids, they start to become more conscientious about their health and are willing to make significant changes to prolong their life. But it's even better if we can start screening earlier for atherosclerosis. Measuring early plaque appearance by age twenty-five will not only help patients avoid atherosclerosis, but will enable them to search for more obscure inherited abnormalities. And it's cheaper and easier than colonoscopies and mammograms, which are already standard for diseases that, though deadly, are far less prevalent than heart disease.

In fact, routine mammograms could also be used to help diagnose heart disease. OBGYNs can see breast arterial calcifications (BAC) on mammograms, though their presence does not represent an increased risk for breast cancer and has often been ignored. A recent analysis of over two hundred mammographies and CT angiographies, however, showed that women whose mammographies showed BAC have an 8.8 percent higher risk of developing cardiovascular disease. Calcium is a result of cooled plaque throughout the arteries, and so its presence is a sign that there is likely plaque elsewhere in the woman's sixty-thousand-mile highway of blood vessels. The study also found that the older a woman was, the younger she was when she began having children, and the higher her blood pressure, the greater chance she had of showing BAC on her mammogram. These already-routine tests could be critical in diagnosing and treating more women before atherosclerosis gains a foothold. And I'd love to see the Pink Ribbon campaign united with the Go Red for Women campaign that uses a red dress as a national symbol to raise awareness about heart disease in women.

So often, we already *are* doing the right tests, but not looking for the right results. OBGYNs and urologists, usually at the forefront of men and women's health, could easily become early defenses on the frontlines against atherosclerosis. Even dentists could contribute to the fight. The panoramic X-rays taken by dentists often show carotid calcifications, a sign that the patient likely already has heart disease. One study found that of seventy-one patients whose dental exams showed carotid artery calcifications, forty-one experienced a major adverse cardiovascular event (MACE) in under three years from the time of the exam.

If an ultrasound of the carotid does show plaque, we need to immediately treat the disease, with no need for further testing to find additional plaque locations. If the disease is in our carotids, it is also in many other sites throughout the sixty thousand miles of vessels. For individuals wanting more complete atherosclerosis imaging, the best next step would be a Coronary Artery Calcium score (CAC) and/or a coronary CT angiogram, both of which can now be done with precise accuracy at extremely low doses of radiation. CAC testing is as widely available as colonoscopies and mammograms, and far less expensive. As atherosclerosis grows in the heart's blood vessels, it begins to calcify, which can be seen on high-speed CT scans. The CAC test measures the total amount of coronary calcium buildup and presents it in a single numerical score, from zero to ten thousand. Any score above zero indicates the presence of atherosclerosis, as there cannot be calcium without plaque. The more calcium, the more plaque, and the higher your score. If you receive a calcium score above zero, you need treatment. You have now seen the disease. In the Boone Heart Protocol, we only need a single calcium score over the lifetime of our patient. This is because, when you treat plaque, it actually becomes more calcified, which indicates healing and stability. So a Boone patient with a calcium score of twenty might have a calcium score of eighty-nine after a year of therapy. This seems counterintuitive, but we are in the business of turning bubbling volcanoes into stable, calcified mountain ranges.

In the more advanced Coronary CT Angiography, dye—called contrast material—is injected into the bloodstream while the patient undergoes a CT scan to illuminate blood clots, growing blockages, and damaged blood vessels throughout the heart and surrounding organs. From this test, doctors can determine the presence of and visualize the plaque in your heart. Recent developments have allowed these images to be further interpreted using artificial intelligence (AI) to identify both the calcified portions of your plaque and the soft, hot plaque throughout your system. Companies like Cleerly, HeartFlow, and Elucid have allowed even deeper interpretation of both old and new CT scans. Coronary CT Angiography with additional AI analysis represents the pinnacle of atherosclerosis imaging in the heart at the time of this writing. Many Boone Heart patients want this sophisticated imaging performed, and it is remarkably impactful and motivating to see the actual plaque lurking in the vessels of your heart. These tests are now used even in hospitals to determine the need for Hail Marys, like stents and bypass surgery, but at Boone Heart, we mostly use these advanced technologies to identify and observe the regression of heart disease over the years.

Researcher Valentin Fuster, MD, has discovered that the first appearance of atherosclerosis in the human appears in the iliac femoral junction of the lower extremities. At Boone Heart, we have used this finding to pioneer the use of ultrasound imaging in the legs, extending from the thigh to the feet, to measure atherosclerotic plaque and blockages in their earliest stages. This requires little advanced technology and no radiation exposure. It is also remarkably accessible to watch regression of disease. When we performed this test (an add-on to a test called Ankle Brachial Index [ABI], which is used to determine who needs amputation due to atherosclerosis), on my wife, Colette, we discovered a 40 percent blockage in her leg. We already knew she had Lp(a), but this was our first discovery of plaque in her regular imaging regimen. We were able to start her on the right medications to lower that leg blockage, but also, more importantly, the leg plaque was just an easily accessible marker of the plaque that likely was occurring throughout the entire sixty thousand miles of her blood vessels. Now, that 40 percent blockage has decreased to 20 percent, and will continue to shrink, right along with any other plaque in her arteries. She'll never be at risk of a heart attack or stroke, thanks to our interventions.

In the future, I imagine imaging will become even more universally accessible and higher tech, as artificial intelligence opens new doors for accurately predicting heart disease.

Why Is It There?

When plaque has been identified, our next task is to determine why it exists.

For some people, the "why" will be obvious: If your father and grandfather and great-aunt all died of heart disease, it's likely coming for you, too. But for others, the *why* may not be as evident, especially if they are young and don't exhibit many other early signs of heart disease. In the old days, our only genetic information about you was your family history. But now, a more advanced genetic screening may reveal any one of the genetic risk factors associated with heart disease—Lp(a), HeFH, MTHFR mutations, Apoe4, Factor V Leiden, Prothrombin (Factor II) mutation—although it's not necessary to get genetic testing unless plaque has been identified. Plaque detection should guide genetic testing and not the other way around. I like to remind my patients that bad genetics just make things appear earlier. But more importantly, every human is going to develop this disease, which is why plaque imaging is always the tip of the spear.

In addition, the presence of other risk factors like high cholesterol levels, high blood pressure, increased blood sugar, insulin resistance, lack of nutritious diet, poor exercise habits, chronic stress, or inflammation may help to explain how plaque has taken hold. Perhaps they've been suffering from rheumatoid arthritis, lupus, diabetes, obesity, cancer, or other related diseases that contribute to damage within the blood vessels. We may discover that someone's lifestyle and/or environment is particularly atherogenic. If they've been smoking cigars for the past ten years, or working the night shift in a high-stress job, it's almost certain that they've got plaque.

Answering the question of "why" will help us create a personalized treatment plan that attacks atherosclerosis from every possible angle of medications, supplements, and lifestyle.

Is It Vulnerable?

It's not enough to know that plaque is there, which we can tell just from the appearance of calcium on imaging. We also have to know whether that plaque is vulnerable; that is, how likely is it that the volcano will erupt? To answer this question, we assess plaque biology, density, vulnerability, and composition. Once the oxidized LDL particles get underneath the lining of the artery and start attracting immune cells to repair the damage, it ignites an inflammatory response that we can track with certain tests. Finding these indicators is much more important than waiting for blockages and symptoms of chest pain. If we continue to wait for these symptoms, it's likely that over half of us will die.

To help us assess the plaque's vulnerability, we can use an assortment of tests. Treadmills with either electrocardiography, stress echocardiography, or nuclear cardiology are the most common tests and may be helpful if you already have chest pain or other cardiac symptoms, but these normal tests can only determine, at best, that you do or do not need to be hospitalized for surgery or vascular intervention.

The better test we use to identify vulnerable plaque is the highly-sensitive C-reactive protein (hs-CRP) test. Elevated levels of hs-CRP in the blood act as a marker of acute inflammation, and any result above 1.0 means you likely have hot plaque in your arteries. The good news is that hs-CRP tests are very affordable and can be done at the same time as other testing, such as a genetic screening. A more targeted test to measure plaque inflammation and vulnerability is the Lp-PLA2 test. This is like a plaque-specific hs-CRP, and I find it to be the most specific test to identify "hot volcanoes."

The MOST Protocol can also help us determine how likely someone is to experience a devastating cardiovascular event. As explained in Chapter 6, atherosclerosis piles the tinder high; stress lights the match that sets it all ablaze. When we evaluate patients using the MOST Protocol, we can help to determine how quickly that match might be lit.

Detecting vulnerability is key in preventing heart attacks and strokes. Remember, most patients will get their cholesterol levels or perhaps a treadmill test if they're lucky, and be told they're fine—when in fact, what kills you is a non-ischemic, non-blood-flow-limiting, non-chest-pain-causing hot plaque that ruptures after a seemingly insignificant stressor. Once it ruptures, a life-threatening clot forms.

Inflammation *is* the alarm bell. The hs-CRP and Lp-PLA2 tests, and MOST Protocol ensure that we heed that alarm.

The next question, then, is: what do we do about it?

What Do We Do About It?

In order to stop heart disease, we have to focus on cooling the inflammation in the body and preventing the oxidation of those LDL particles throughout the bloodstream. But doing so is easier said than done. It requires a multifaceted approach, combining the right medications with lifestyle changes and stress management.

Medications

Most patients (and most doctors) start with a focus on lifestyle changes: "Stop eating cheeseburgers every day, and you'll be fine." That advice, plus the skepticism and reticence many people feel against taking pharmaceutical drugs of any kind, prevents too many people from getting lifesaving medications that could drastically reduce their chances of a heart attack. Lifestyle alone is not enough to prevent heart disease when you've got genetics, environment, and stress undermining your efforts at every turn. The most effective way to prevent further progression of disease *and* reverse the plaque that's already there is to take supplements and medications personalized for your genetics and health needs.

For most people, this means starting with a simple aspirin every day. This isn't going to cure the disease, but it ensures that if a fracture in the plaque occurs—that is, if the volcano erupts—a life-threatening clot isn't going to form. In our battery of tests, we measure 11-dh-TxB2/Cr ratio, which defines the need for and the effect of aspirin on an individual. The test is trademarked as AspirinWorks, and is a simple test of the urine. The test identifies people whose blood clots particularly quickly, which was likely a benefit for some of our ancestors during more violent times. Aspirin might be the most dangerous drug I prescribe my patients, because it can cause ulcers and bleeding for those with certain pre-existing conditions, like gout, or kidney or liver problems. However, it can be lifesaving if you have plaque and inflammation that might lead to plaque rupture. Never start a new medication without first consulting with your doctor to ensure that the use of aspirin won't put you at further risk of harm. But most doctors will tell you not to take aspirin until *after* you have a heart attack, and that's simply far too late. There has been a lot of controversy over the last century about who should take an aspirin to prevent heart disease. We believe the Boone Heart Protocol solves this controversy by using aspirin only in those who have been precisely identified with plaque and elevated AspirinWorks result. Why wait until there's a problem, when you could prevent the problem now?

Statins

The most common medications in the treatment of heart disease are statins. Statins are a type of medication, also found in red yeast rice, that is designed to reduce the LDL cholesterol in the bloodstream. Statins include many brand names that you've probably heard of or seen on commercials, particularly Lipitor® and Crestor®. A 2022 meta-analysis of studies assessing the efficacy of statins found that statins significantly reduce the risk of cardiovascular events, including myocardial infarction (heart attack) and stroke. As with any medications, statins have some side effects, including benign liver damage for those at risk. However, the same meta-analysis found very little evidence of increased risk for harm due to the use of statins. Additionally, studies from the Cholesterol Treatment Trialists' Collaboration have found that "Reduction of LDL cholesterol using statin therapy substantially reduces the risk of major vascular events (major coronary events, strokes, or the need for coronary revascularization*)* and vascular mortality by about one fifth for each 1 mmol/L reduction in LDL cholesterol achieved." In layman's terms, that means that, for every 38 points your LDL cholesterol decreases, you reduce your risk by 20 percent.

The research seems clear that for the majority of patients, statins represent the best way to stabilize atherosclerotic heart disease and prevent it from getting any worse.

Interestingly, because statins are also anti-inflammatory, they are becoming a more prevalent drug in treatments for diseases beyond CVD. Many doctors observed that the use of statins greatly lowered patients' mortality risk during the COVID-19 pandemic. Researchers in Sweden confirmed that "Statins could confer protective effects among COVID-19 patients by stabilizing and promoting regress of atherosclerotic plaques along with anti-inflammatory effects." I'd seen the same positive impact of statins on patients with AIDS, so at the Boone Heart Institute, we prescribed statins for every COVID patient. Unsurprisingly, not a single one of our patients died due to COVID-19, or even had to go to the hospital for treatment. Additionally, statins are starting to be prescribed to breast cancer patients, as they've been shown to lower the risk of mortality and recurrence.

Despite these positive results and the growing number of studies proving their widespread efficacy, I still encounter skepticism and distaste amongst patients who would prefer to rely solely on organic foods and exercise to fight their heart disease. This attitude seems to be prevalent in places like Boulder, Aspen, Santa Fe, and California. Those folks will take any drug as long as it's not made by big pharma. In places where people tend to pride themselves on their healthy lifestyle, nobody wants to take a statin. This is yet another one of the dangerous effects of a health industry that tells people that if they only make healthy choices, they'll be able to avoid all kinds of disease, regardless of their genetics, stress levels, and socio-economic context. When they get the disease anyway, they worry that it means they didn't do a good enough job or make healthy enough choices, and if they take a statin, other people will know they failed. Women, particularly, worry about the social stigma surrounding heart disease as it relates to body weight, and tend not to get the help they need early enough. I tell people: not smart.

A 2021 study from Denmark found that people who stopped taking statins were more likely to experience a major adverse cardiovascular event (MACE) in the five years after stopping their prescription than those who continued taking their statins. One reason I hear patients complain about statins is that it makes their bones and joints hurt. But as an old guy myself, my bones and joints hurt all the time anyway. I could easily say I'm not going to take a statin, but I've gone off my statin before, and I still had all the same aches and pains. The incredible benefits of staying on statins simply outweigh the minor side effects.

Much like ultrasound imaging and hs-CRP testing, statins are widely available and have become much more affordable in recent years. After the American College of Cardiology and American Heart Association updated their guidelines in 2013 to broaden their recommendations for statin usage to more patients, the number of people taking statins dramatically increased. In 2008-2009, about thirty-one million people reported taking statins; ten years later, that number was ninety-two million. As a result, the out-of-pocket cost of statins also decreased, from $348 a month at its peak in 2010 to $30 in 2019. Now, the majority of statin prescriptions cost patients between $9 and $20 per month.

Ninety-two million people sounds like a lot of people taking statins, and it is—but remember that many of those people are taking statins for other conditions, and there are roughly 128 million Americans living with diagnosed heart disease, and likely another 100 million that are undiagnosed, according to Boone Heart projections. Progress is being made, but there is still a long way to go.

PCSK9 Inhibitors

While statins are by far the best first line of defense, for those who have extensive plaque, a statin alone may not make enough of a difference in the amount of LDL cholesterol in their bloodstream to prevent further oxidation and inflammation. Others may have cholesterol that proves to be resistant to statins. For those, the ideal intervention is a prescription for evolocumab, known by its brand name Repatha®, or alirocumab, known as Praluent®. These drugs are usually taken via injection every two to four weeks. Another PCSK9 small-interfering RNA (siRNA) therapy is inclisiran, known as Leqvio®, which is injected just once every six months. These medications work by inhibiting the function of PCSK9, a gene that controls the breakdown of LDL receptors in our liver. Researchers discovered that patients who have certain mutations of their PCSK9 gene experienced lower concentrations of LDL particles in the bloodstream, and therefore were at lower risk of atherosclerosis. By blocking the function of PCSK9, we can prevent the breakdown of our LDL receptors—thereby allowing the liver to continue effectively clearing out the excess LDL. Remember, cholesterol is a good thing, as long as it is not floating around in excess inside the bloodstream.

The results of using these PCSK9 inhibitors have been astounding, and they have been the single most significant factor in improving outcomes for my patients. A 2022 study of patients with a history of heart attack who were treated with evolocumab found that not only did patients' levels of LDL cholesterol decrease, but the medication also increased the thickness of the roof of the plaque—meaning, the plaque became more stable and less vulnerable to rupture. In "cooling" this hot lava plaque, the lipid arc of the plaque also decreases, as shown in Figure 9.1. This is consistent with other findings that the thicker the fibrous cap (roof) on atherosclerotic plaque, the less likely it is to rupture and cause an acute cardiovascular event. On the flip side, a thin fibrous cap with even a moderate narrowing of the blood vessel—even as low as 30 percent!—indicates a major risk for plaque rupture. Repatha has been critical in helping thousands of patients avoid MACEs.

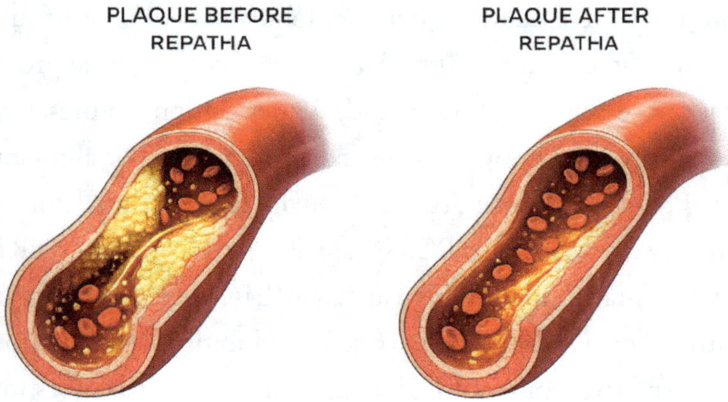

Figure 9.1

Cardiologist Marc Bonaca studies peripheral artery disease (PAD), the narrowing of arteries that carry blood from the heart to the legs. PAD is common in those with diabetes and other related diseases, and can often result in lower limb amputation. In a 2018 study, Bonaca and his team found that taking evolocumab significantly reduced the risk of both amputation and MACEs for PAD patients. Results like this are groundbreaking, and the findings still aren't widely known. I recently went to a three-day conference in Chicago with the world's best vascular surgeons and imagers. We spent the bulk of those three days talking about all the imaging that was necessary to find and treat blocked arteries. Presenters showed all kinds of horrible pictures of amputations of legs. Meanwhile, there was not one mention of even a statin, let alone Repatha, both of which have been shown to *reverse* and prevent the disease. It was a poignant reminder that most of these doctors are paid to salvage the limbs of these ravaged patients, who are often suffering in food and healthcare deserts, unsupported by a medical system that waits until the last possible moment to intervene.

So why doesn't everyone take Repatha? If I had my way, nearly everyone would be, and no one would ever have a heart attack. In fact, PCSK9 inhibitors represent the only class of drugs proven to actually reverse the growth of plaque by thickening the fibrous cap, when added to baseline statin therapy. However, even for those who know they have heart disease *and* their doctor recommends Repatha® (beyond just statins), the cost may be prohibitive. At roughly $560 per month before insurance, Repatha® may be feasible for those in the upper middle class who are dedicated to their health, but is still inaccessible for many Americans. Fortunately, many insurance plans and Medicare cover Repatha®, and I expect we'll see the price decrease in the coming years, just as we saw with statins, to allow more patients access to this lifesaving drug.

GLP-1 Agonists

On March 18, 2024, Oprah hosted a prime-time special dedicated to a controversial topic: weight-loss drugs, particularly semaglutide (Ozempic® and Wegovy®) and tirzepatide (Mounjaro® and Zepbound®). In the event, Oprah joined dozens of celebrities who have been vocal about their use of this new class of drugs, which has seen tremendous results in helping people lose weight and manage diabetes. These drugs, usually taken weekly via injection, are called glucagon-like peptide-1 (GLP-1) agonists, which reduce blood sugar by stimulating the production of insulin and decreasing appetite. Recent studies have shown that semaglutide not only helps in the management of diabetes, but also reduces the likelihood of MACEs for those with obesity but without diabetes.

The reaction to these medications has been even more pronounced than the hesitations around statins and PCSK9 inhibitors because of their prevalence as weight-loss aids. In my mind, however, their biggest benefit is the decreased risk of heart disease and the management of diabetes, both of which can be devastating for a person's health; the weight loss is a nice bonus. Moreover, these drugs are extremely safe. At the Boone Heart Institute, we've had five hundred patients on Ozempic with absolutely no adverse side effects. Right now, they are expensive (Ozempic costs an average of $900 a month before insurance, and insurance only covers this prescription for those who have been diagnosed with diabetes), but as more awareness of these drugs and their health benefits builds, the prices will decrease as well.

Vitamins and Supplements

In the mid-2000s, a group led by Kedar Prasad, Ph.D., America's first Ph.D. in radiation biology, worked with researchers at NASA and the Department of Defense to develop a proprietary micronutrient and antioxidant vitamin blend to help defend astronauts and soldiers against radiation exposure while on the job. I found Dr. Prasad's group, then called Premier Micronutrient Corporation, in my search for a way to protect my patients from the ionizing radiation exposure that comes with CT imaging. There's only so much outer protection astronauts and soldiers can wear; they need additional protection that works from the inside out. I wanted the same for my patients.

A study in Germany using the PMC vitamins showed that individuals taking these micronutrients suffered 57 percent less double-strand DNA breakage following CT scans than those not taking protective antioxidants prior to their scan. I began giving these vitamins to my patients before and after their CT imaging. Amazingly, more protective benefits of this micronutrient formulation began to emerge. We found in studies of these vitamins that they decreased tinnitus (ringing in the ears), a common hearing condition suffered by soldiers who are exposed to loud noises, but more importantly, we saw that people who took the vitamin formulation, now called Multimmunity, experienced less DNA damage after being exposed to radiation, such as a CT scan. Another study out of the Naval Medical Center in San Diego revealed that marines taking these vitamins recovered 50 percent faster after having experienced nonpenetrating traumatic brain injuries from a blast. Senator Mark Kelly, who used to be an astronaut, has taken these vitamins and recommended them for his wife, former U.S. Representative Gabby Gifford, after she was shot at point-blank range in the head in 2011, to aid in her recovery.

From 2008 to 2009, Boone Heart Institute ran a study called the MAVRIC (Micronutrient Antioxidant Vitamin Reversal of Intima media thickness of the Carotid artery) trial, in which healthy firefighters were given Multimmunity vitamins twice a day for one year. We found that they experienced a 15.7 percent decrease in LDL cholesterol, a 20.8 percent increase in HDL cholesterol, and a 30.3 percent improvement in the LDL-C/HDL-C ratio. Additionally, the vitamins were proven to slow the typical age-related accumulation of plaque within the arteries.

There are many excellent immunity-boosting vitamins available on the market. We always recommend that our patients take one as the first line of defense to reducing their inflammation and oxidation. Doing so will also help to fill in the gaps in our nutrition (nearly everyone has some kind of vitamin deficiency) to prevent cancer and other diseases of aging as well. I wouldn't bet my life on just a vitamin, but starting to take a high-quality vitamin early will give you a massive head start on beating disease.

Depending on your unique medical history, genetic profile, and other factors, doctors may recommend a number of other medications or supplements as well, such as bempedoic acid, ezetimibe, omega 3s, new oral anticoagulants (NOACs), herbs, blood pressure medications, blood sugar medications, sodium-glucose transport protein 2 (SGLT2) inhibitors, metformin, or other inflammation-controlling medications. Your doctor will be able to advise you on the best options for your health.

There's no doubt that medications are the best and first line of defense against heart disease. They represent our best shot for lowering the toxic amounts of LDL cholesterol running through most of our bloodstreams and preventing the buildup of atherosclerotic plaque in our arteries. However, we shouldn't rely only on medications to accomplish our goals; medications work best when they are accompanied by lifestyle changes that help prevent the disease from getting worse.

Proactive Lifestyle Changes

In my practice, I don't spend a lot of time lecturing my patients on healthy lifestyle choices, because most of us already know the changes that we should be making. However, it's worth elaborating on a handful of basic principles that work for most clients. Of course, any changes to your diet and exercise should always be undertaken with caution and with the advisement of your doctor.

Nutrition

Before I became a physician, I was still something of an expert in fitness and health because of my master's degree in exercise physiology and my work at the YMCA. I tried to practice what I preached in this area. So when we took my son, Tyler, to McDonald's, he was only allowed to eat the healthiest option on the menu: the Filet-O-Fish. Because fish, of course, is healthy! In those days, though, they didn't publish nutritional information on menus, so we had no idea that we were feeding our child one of the most high-calorie, high-fat, unhealthy items on the entire menu. A burger or chicken nuggets would have probably been better! Once we realized the nutritional facts, we changed course. I did not keep feeding my kid the sandwich and insisting it was healthy just because "everyone knows fish is healthy."

The point is: when we know better, we can do better.

Now, fortunately, most menus make their nutritional facts publicly available so that we can make informed decisions no matter where we go. Plus, our general awareness about nutrition has come a long way from the "food pyramid" many of us were taught about in previous decades. For me and my patients, I boil nutritional counseling down to a handful of simple choices that work for most patients: whole grains, healthy fats, low quantities of meats, lots of veggies and fruits, and nothing processed. For heart health, I often recommend a Mediterranean diet and even WeightWatchers. I like to caution patients about intermittent fasting, because the research on time-restricted eating is limited and the results have been mixed, but some studies suggest a possible increased risk of heart disease for those who practice intermittent fasting. However, I have seen this approach help many in the pursuit of high-performance longevity and weight loss. Ultimately, any diet that restricts the total excess caloric intake that comes with the modern American lifestyle is a good starting place. Medicines like Ozempic and Monjauro have proven to be a game changer for those who have had lifelong troubles with diet and weight; they are incredibly safe and confer positive effects throughout the body's multiple organ systems.

Exercise

You already know that excessive exercise is incredibly risky for the heart. Remember Dr. David C. Nieman's model of the J-shaped curve, which shows that moderate exercise has innumerable health benefits, including lowering our inflammation and blood pressure (thereby decreasing our risk of Type 2 diabetes), and increasing the strength and capacity of our lungs, heart, muscles, and bones.

Moderate exercise also has been shown to help with stress regulation. In one study, those who regularly exercised were shown to be more resilient and suffer fewer emotional declines when subjected to an acute mental stress test. The American Heart Association recommends exercising about thirty minutes five times per week (for a total of 150 minutes), with moderate to high-intensity weight-bearing exercise at least twice per week to gain as much benefit from exercise as possible. Brisk walking, swimming, dancing, sports like tennis, biking, hiking, and running are all good options for getting the recommended amount of movement while having fun.

Stress Regulation

As you know now, regulating stress is one of the most important things you can do to protect your heart, no matter whether you have atherosclerosis or not. There are several regular practices for managing stress that we often recommend patients take up, especially if we find during the MOST protocol that their bodies experience more severe reactions to mental stress. One of the most simple and accessible stress-reduction interventions you can practice is regulated breathing. However, a 2023 review of the studies around breathing practices found that not all are created equal; the best ones avoid fast breathing paces, last no less than five minutes, and incorporate regular instruction from a qualified teacher over multiple sessions. Other mindful practices like meditating, spending time in nature, and practicing yoga have all been found to have a positive impact on mental health while protecting against the harmful effects of stress. If recommended by a doctor or psychiatrist, Selective Serotonin Reuptake Inhibitors (SSRIs) may also be helpful in moderating the impact of anxiety and depression.

The Proof Is in the Results

Years ago, I spoke to a group of corporate leaders in New Mexico, most of whom were executives at oil companies. One of the men there was Brad Bartek, the CFO at one of those companies. Brad liked what he heard at my lecture and arranged a meeting with the founder and CEO of his company, an elderly gentleman named Mack Chase, who was suffering from some known heart conditions.

Mack was an old-fashioned patriarch who used to be so poor he couldn't buy his kids shoes. After discovering oil when he was a young man, he'd become a magnate and real estate tycoon, buying up ranches throughout the state. By the time I met him, he was the wealthiest person in New Mexico. Suffice it to say, he wasn't easily impressed. Mack flew into the private airport at Denver, where I was instructed to meet him on the tarmac. Brad told me, "He's going to talk to you for a few seconds, and if he likes you, we'll go to your clinic. If not, we're flying back to New Mexico."

I did as instructed and joined Mack in his car on the tarmac. We chatted for a few minutes, and then he said, "Well, let's go back to your office."

At that point, Mack was seventy-five and had already had a stent. In fact, we could see in his imaging that the plaque in his arteries was so advanced it was a miracle he hadn't already died. Fortunately, we started him immediately on an aggressive regimen of medications as part of the Boone Heart Protocol. His health and vitality increased immensely over the next decade, and he was able to oversee the succession of his empire to his children and grandchildren with wisdom and peace. Mack lived to the age of ninety-two, with no further surgeries needed.

After that, word got around in the industry and we began to see a lot of oil executives who told us just to do "whatever we'd done with Mack."

One of those people we subsequently treated was Brad Bartek. He had flown to that meeting with Mack, as a healthy sidekick trying to convince his boss to take better care of himself. We convinced Brad to join Mack in getting some preliminary testing, despite his protests. Little did he know what we would discover.

Brad was known in the oil industry as a bit of a risk-taker. Famously, Mack had hired him when, in considering whether to make a major investment or not, Brad decided the matter with a flip of a coin. Mack liked that kind of daring, and Brad liked that we were on the cutting edge of cardiovascular disease. When we first saw him, Brad was fifty-two, and we discovered a 71 percent blockage of soft, vulnerable plaque on his CT angiogram, which was a brand-new technology at the time. With the results of the MOST Protocol and his adrenaline-seeking business practices, we knew he was at grave risk of a heart attack. He wasn't experiencing chest pains and he did all right on the treadmill test, but I knew that if he went to any other cardiologist, the standard of care would be for him to have a stent placed immediately.

I had another idea.

I told him, "Everything we're going to do is going to stabilize that plaque. If you want to go back to Dallas, they're going to put a stent in. But I wouldn't let them do it to me."

Brad agreed, and over the next sixteen years we saw his widowmaker blockage decrease from 71 percent to 39 percent, not only treating the volcano that was building in his artery, but effectively protecting all sixty thousand miles of his blood vessels. At that point, he had been on Crestor®, or rosuvastatin, and it had gotten him far. But in 2016, we started him on Repatha®, and his plaque decreased even further to 28 percent. It's been over fifteen years since we first started treating him, and he's now approaching retirement, but I imagine he'll keep tossing coins and wheeling and dealing for a long while yet.

At the Boone Heart Institute, we've seen hundreds of patients just like Mack and Brad. We know that the medications and protocol we recommend works. It's deceptively simple, and the medications we recommend are accessible to far more people than take advantage of them. And yet, we've been fighting an uphill battle for decades to convince more doctors and institutions to adopt these practices and recommendations. Institutional resistance is hard to overcome. However, with the body of research supporting our efforts growing every day, I remain confident that we'll see a new mindset around heart disease prevention take hold and millions of lives will be saved as we eliminate CVD as the world's number-one killer.

About ten years ago, I woke up one night in a cold sweat when I had the sudden, clear realization that even though we could save our patients from heart disease, many of them would just end up dying from cancer or dementia instead. Could we do something about that?

I had seen the ability of the Boone Heart Protocol to save lives. If our patients follow our recommendations, it's nearly impossible for them to have a stroke, heart attack, or other life-threatening event. I've even worked with several patients who have cancer, who have told me that their doctor said, "We're worried about your cancer, but you're much more likely to die of heart attack and stroke before the cancer kills you." Excluding heart disease, which still ranks as number-one, these other "diseases of aging" are by far the biggest threats to human life. I started to look into this problem and have been excited to learn about the presence of a growing body of research that supports the idea that the interventions and medications we recommend as part of the Boone Heart Protocol aren't just curing heart disease, they're also helping us all live longer, healthier lives free from cancer, cognitive decline, and other diseases as well.

CHAPTER 10

Healing from Head to Toe

Many years ago, a young boy in Colorado snuck a kiss with one of the girls in his eighth-grade class. Soon after, his family moved across the country, and he started attending an elite private school in New Jersey. He excelled in sports, playing lacrosse, football, and several others. He went on to play lacrosse for the Army at West Point, which has consistently been rated one of the top lacrosse teams in the country. Meanwhile, the girl he had kissed had also started playing lacrosse, leading her team to victory in the state championships for multiple years before going on to star for the University of Denver women's lacrosse team in multiple NCAA tournaments.

After college, the two continued on their separate ways. The boy served in the Army for several years and then moved back to Denver for a career in finance, while the girl began coaching high school lacrosse while studying for a master's degree in commercial real estate. But fate conspired to bring them together again. The boy's younger sister was also a lacrosse player, and her coach happened to be the girl he had kissed in eighth grade. "You have to meet her," his sister said.

The girl was my daughter, Kiki, and the boy was my soon-to-be son-in-law, Ehret. And if I'd known about the kiss in the eighth grade, I wouldn't have been too happy about it. Fortunately for him, I didn't find out about that until much later.

Ehret remembered Kiki and called her up to ask her out. She was dating someone else at the time, but her boyfriend agreed it was fine—just two old friends catching up, or so he thought—which, in retrospect, was a pretty dumb move on his part. Over dinner and drinks, they rekindled their early romance and were soon dating.

Now, you can't be connected to my family for too long before you find yourself in my office for some kind of testing, and I knew exactly what test I wanted to give to the man courting my only daughter.

We had Ehret put on a futuristic-looking headset as part of a sophisticated brain scan called WAVi. The WAVi test measures brain performance using electroencephalography technology to measure brain speed, brain power, and reaction time. We found out that Ehret's brain power was about 6 millivolts, approximately the brain speed you'd expect to find in an eighty-year-old. For reference, a typical voltage for someone his age would be 12 to 15 mv. In Ehret's defense, these tests don't measure a person's intelligence. Ehret was plenty smart, but his experience in the Army and with lacrosse hadn't done his brain any favors. Still, there was no way I was going to let a guy with 6 millivolts of brain power marry my daughter.

Through his genetic testing, we also learned that Ehret has the ApoE4 genotype (the same type that Chris Hemsworth discovered he had on his television show), which could have been a contributor to his brain's lower power at that stage in his life, along with his multiple concussions from playing high-level contact sports. The ApoE gene is associated with the brain's plasticity and ability to repair. Studies have shown that children and adults with the ApoE4 genotype have a higher incidence of poor outcomes after traumatic brain injuries (TBIs).

Because he was young and otherwise quite healthy, we started Ehret on the Multimmunity vitamins as a first-step intervention to healing his brain. Just six months later, without any other interventions, his voltage measured 16 mv, high enough that I happily gave him my blessing to marry Kiki. They were married in a beautiful ceremony in Denver, and now have three wonderful children. I feel better every day knowing that my daughter and grandchildren can expect to enjoy a long life with Ehret, armed as he is with knowledge about his health and how to preserve his brain's power for as long as possible. I'm also glad that Kiki and Ehret can make the best parenting decisions for their children and educate them on taking care of their health, knowing they may have inherited the ApoE4 genotype as well.

What's more, using the same simple interventions, we may be saving Ehret, Kiki, my grandchildren, and all of my patients from other diseases and problems. Heart disease, stroke, dementia, cancer, and other diseases all stem from the same sources: aging, oxidation, and inflammation. Your entire body, from your head to toes, is connected through the cardiovascular system, and every part is fed by the same network of blood vessels. Heart disease *is* blood vessel disease. This is why curing heart disease holds the potential for greater healing than we may previously have realized.

Heal the blood vessels, and you heal the entire body.

Preventing Cancer

In 2019, I was lecturing at a conference in Arizona where I heard a presentation that changed the way I thought about heart disease and cancer forever. The presenter was William Li, author of the *New York Times* bestselling book *Eat to Beat Disease*. Li is a scientist and internal medicine specialist who has studied extensively the role of angiogenesis, the body's process for creating new blood vessels from endothelial cells (sprouting angiogenesis), or splitting an existing blood vessel to create new pathways (splitting angiogenesis). Both types of angiogenesis are essential in times of rapid, intense growth, such as the development of a placenta and embryo, the healing of a wound, or recovering after a workout. And, it turns out, in the spread of cancer cells.

It may surprise you to learn that everybody has cancer cells in them all the time. Young and healthy bodies usually fight them off, but stress, lifestyle, aging, and genetics can hinder the body's ability to battle these foreign cells. If the body cannot fight them off, those cancer cells may form together to create a tumor. Just like muscles that need a constant supply of oxygen and nutrients to continue functioning, tumors need oxygen and resources to spread throughout the body, too. Even more cunning, if cancer cells aren't receiving enough oxygen and supplies to continue their spread, they send out biochemical signals to stimulate angiogenesis to ensure that they get those resources, just as our healthy cells do. "Worse yet," Li writes, "the same blood vessels that feed cancer tumors also serve as exit channels for malignant cells to escape into the bloodstream." In other words, through angiogenesis, cancer cells not only grow, but also spread to new parts of the body.

When I met him, Li was developing chemotherapy agents, with the idea that instead of stopping the army that's trying to invade you and kill you, you cut off their supply lines. It's a very common tactic that has been used in military operations from the Middle Ages to the Civil War and World War II: if you cut off the enemy's supply of water and food, they die. Li's research has centered around how certain foods (think processed foods, French fries, and doughnuts) can "feed" cancers that take advantage of our natural process of angiogenesis, while other foods (like many fresh fruits and vegetables) can "starve" cancers.

Dr. Li and I heard each other's presentations and immediately saw the connections between what we were doing. I'm trying to prevent those blood vessels from developing plaque, and he's trying to prevent them from being co-opted by cancer. And it turns out, many of the techniques used to heal one also heal the other. Initially, Li evaluated components of chemotherapy to determine their effect on angiogenesis. Then he looked at food products through the same lens. Eventually, he started using that evaluative process on cardiovascular drugs. And he found that the drugs that I liked, such as ACE (angiotensin-converting enzyme) inhibitors, blood pressure medications, and statins, all tend to protect the heart and the brain while starving cancers. This means that all the cardiovascular drugs we're talking about to save your heart are also probably equipping your body to fight off cancer cells.

These drugs are not actually cancer therapies, but in many cases, they are being given alongside chemotherapy and other cancer treatments to help the body right that process of angiogenesis that has gone wrong. In the last chapter, I mentioned the widespread use of statins, which have an anti-inflammatory effect, to aid the effectiveness of chemotherapy for women undergoing treatment for breast cancer. A 2019 study of over twenty-nine thousand participants also showed that patients already taking a statin at the time of diagnosis with colorectal cancer have an increased chance of survival. Statin use may vary in its effects on cancer and cancer treatments, depending on the type of cancer and specific treatment, so it's important to talk to your doctor if you have questions about the possible benefits of statins if you or someone you know is undergoing treatment for cancer.

New studies have also shown that PCSK9 inhibitors can be powerful agents in boosting the efficacy of cancer treatments. In particular, patients undergoing treatment for cancer with immune checkpoint inhibitors can be at greater risk of developing atherosclerosis. PCSK9 inhibitors have been shown to not only help prevent atherosclerosis in these patients, but also make the cancer cells more susceptible to the treatments.

Over the years, I've come to believe that the strategies we recommend for taking care of your heart are also the best strategies for preventing cancer. All the risk factors are the same: If you have diabetes, obesity, high blood pressure, inflammation, C-reactive protein, and high cholesterol, you're more likely to develop atherosclerosis *and* cancer. This is significant, because despite institutional resistance to basic interventions like imaging, there is a lot that we can measure in the heart and blood vessels to detect the beginnings of atherosclerosis. It's much harder to measure cancer in its initial stages, because you don't know where to look. Typically, the average U.S. adult will only get screened for a handful of possible cancers in their lifetime: breast cancer and cervical cancer for women, prostate cancer for men, melanomas, and colorectal cancer. And that's if they are proactive about getting screened. And yet, these cancers only make up about 40 percent of cancer diagnoses. The other 60 percent of cancer diagnoses occur in organs and areas of the body that we do not screen.

This is a problem that many researchers and biomedical companies are trying to solve with multi-cancer early detection (MCED) testing, and again, the answer is found in the blood. One such test, the Grail Galleri® test, analyzes a blood sample to detect the possible presence of over fifty types of cancer and where it might be located in the body. It's like a biopsy of the blood, looking for specific biomarkers for cancer that might be present in cell-free DNA. Because these tests rely on blood, they are most effective for detecting cancers that spread quickly like lymphoma, leukemia, and ovarian, pancreatic, or gallbladder cancer.

In just the last two years, we've had about a hundred patients on the Boone Heart Protocol take the Galleri® test and identified three with early-stage cancers. For one patient, the Galleri® test helped to pinpoint that the cancer was in the upper esophageal tract from his mouth to stomach. We were able to refer him to a gastroenterologist and continue supporting him during his treatments. He's fully recovered now from esophageal cancer and doing well. Another one of those patients is a young woman who's just recently been diagnosed with fast-growing ovarian cancer. Thankfully, the Galleri® test caught it in the earliest stages and her prognosis is excellent because it was found so early.

I like to imagine that in twenty years MCED testing like this will be commonplace, and by then the technology will be even better and more precise to help us detect cancer early on. At the very least, I hope patients like you will feel empowered to request tests like this until they become standard practice by primary care physicians. In the meantime, it's clear that by protecting your heart and blood vessels, you can help prevent cancers that might not be found until something goes drastically wrong.

But taking care of the cardiovascular system has implications beyond just cancer. If keeping blood vessels healthy is a fundamental key to stopping the growth and spread of cancer, what else could we prevent?

Preserving the Brain

In 2009, I was working with the NFL Alumni Program that Boone Heart pioneered, testing retired players for signs of heart disease and educating them on the power of certain vitamins and medications. After one event in Pittsburgh for retired Steelers, we held a lecture with Dr. Joseph Maroon, the Steelers' neurosurgeon. In the audience that day were famous Pittsburgh Hall of Fame players like Charles "Mean Joe" Greene, L. C. Greenwood, Jack Lambert, and Lynn Swann. Many of the players there had experienced repeated minor head traumas and concussions from their time playing; the impact of how this trauma manifested later in their lives was still largely unknown. Dr. Maroon was most concerned with helping the former players safeguard their brains for the rest of their lives. On the stage, he told them,

> "We're certainly working on better helmets. We certainly have developed better concussion protocols. You all loved playing in the NFL. And you'd probably still do it again, even. It doesn't take a genius to show that banging your head against the wall is probably not good for the brain. But the most important thing you can do for your aging brain, your formerly concussed brain, and your traumatic brain injury is everything Dr. Boone and his crew talk about."

Dr. Maroon's message followed us to our next NFL Alumni screening with the Green Bay Packers in Wisconsin, all the way to Bart Starr, Jim Taylor, and Dave Robinson, some of the country's greatest football players. It was an honor to be there. Years later, Joseph Maroon's colleague, Dr. Bennet Omalu, discovered that many former

NFL players were suffering from strange symptoms of dizziness, mood swings, forgetfulness, and impulsive behavior, which Omalu finally diagnosed as chronic traumatic encephalopathy (CTE). Dr. Maroon played a pivotal role in understanding the origins of CTE and was one of the doctors depicted in the 2015 film *Concussion* starring Will Smith.

Dr. Maroon's support was a critical moment for me in understanding that the implications for curing heart disease are even more widespread than we previously thought. Just as cancer and every other part of the body is nourished through blood vessels, so is the brain. We already know that those with the ApoE4 genotype, like my son-in-law Ehret, are at greater risk of both heart disease and dementia. But the connections between heart health and brain health go much further than that.

My path to an interest in brain health started with our protocols for radiation protection during CT heart imaging. As mentioned earlier, I didn't want to be cavalier in ordering these tests because of the radiation exposure, regardless of their potential to help diagnose CVD in its early stages. That's why I sought out the Multimmunity vitamins that were developed to protect astronauts, soldiers, and eventually my patients from radiation exposure. Around this time, I met a brain-health pioneer named David Oakley, PhD, creator of the WAVi test, which was developed in Boulder, Colorado, and was being used to measure football players' brains after concussions. WAVi is an EEG, much like an EKG, but for the brain, which allows us to measure patients' brain speed, brain power (voltage), and reaction time. The development of this technology was a landmark moment in concussion treatment, because for the first time, neurologists had a way to measure the impact of those concussions. Unsurprisingly, they found right away that brain speed and brain power got worse after a traumatic brain injury. But there was no clear way to treat those players; the protocol was simply to wait for the concussion to clear and for time to do its healing work. After a couple of weeks, most players' WAVi scan returned to normal, pre-concussion levels, and they went back onto the field. However, some brains never recovered, and probably should never have played contact sports again. Years later, all of those players are more likely to begin suffering from symptoms of CTE.

From my perspective, WAVi had multiple applications in my clinic. First, I knew that my patients—at that time, primarily aging Baby Boomers—would like to have their brain power, speed, and reaction time quantified to see potential decline. Second, I suspected that if younger healthy brains could recover on their own after being damaged, perhaps we could repair my patients' aging brains before dementia or age-related decline as well. WAVi gave me a tool to measure that. WAVi EEG measures of p300 latency and voltage have long been correlated with mild cognitive decline and dementia.

At the Boone Heart Institute, I started testing my patients and myself with WAVi. The early results confirmed my growing suspicions. I already knew that these patients had good heart numbers, good carotid numbers, good abdominal numbers, and good blood numbers, and I found that they scored in excellent ranges on the WAVi test as well, relative to their ages. We tested more patients who had not yet started the Boone Heart Protocol and saw much worse initial results. But over the course of the next year, we saw that their brains improved right along with the rest of their cardiovascular system when treated with the Boone Heart Protocol.

Now, further research is validating and confirming the personal transformations we've seen, particularly with dementia and specifically Alzheimer's. Dementia is an umbrella term that includes Alzheimer's disease, vascular dementia, and other disorders characterized by cognitive decline and memory loss. The common denominator in many of these cerebrovascular diseases is impaired blood flow to the brain, whether by the formation of plaques from the accumulation of amyloid-ß (Aß) peptides (as in the case of Alzheimer's disease) or by damage or clogging of the brain's blood vessels (as with vascular dementia). Angiogenesis "gone awry" is also now thought to be involved in the development of Alzheimer's and other forms of dementia, where the body is failing to create new blood vessels to the brain when needed. There are many unique forms of dementia, but every dementia has a substantial shared component of vascular disease.

My observations on the power of healing heart disease to heal the brain as well began in the late 2000s, before PCSK9 inhibitors were available on the market, when often the only intervention was lifestyle changes and the introduction of vitamins and/or statins. But now that PCSK9 inhibitors are widely available, researchers are beginning to realize their potential not just as an accompaniment to cancer treatment, but also in healing the brain. For example, a 2022 study found that PCSK9 inhibitors increase "cerebral clearance," helping the brain to remove the excess amyloid-ß (Aß) peptides that tend to build up in the brain's blood vessels and form blood-limiting plaques.

There is much debate about the role of PCSK9 inhibitors and their impact on cognitive function. Cholesterol is necessary for the brain to function and continue building new neural pathways, and many doctors worry that because these drugs drastically lower the amount of LDL cholesterol in the blood, brain function can be impaired. Keep in mind, however, that babies, whose brains are more active building new neural pathways and whose bodies are growing at a rate much faster than any adult's, have typical LDL cholesterol levels of about 20 to 25 mg/dl in the womb. After birth, their LDL cholesterol levels triple or quadruple over the next several years to about 75 to 80 mg/dl at age twelve. After that, our body slows down its growth rate and doesn't need as much. And yet, the typical recommendations for "normal" or "healthy" LDL cholesterol levels for adults are anything under 100! However, over the past couple of decades the recommendations for optimal LDL-C have continued to get lower and lower, with new studies showing that LDL-C as low as 40 or even 30 mg/dl are perfectly safe and yield even greater benefits. My belief is that every particle of cholesterol is inflammatory in the bloodstream and needs to be transported as quickly as possible into the liver or the brain where it can be put to work. This can be accomplished now with PCSK9 inhibitors, where I regularly see patients' LDL-C dropping below 30 mg/dl while they experience great benefits to their physical and mental health.

In recent years, I've seen more evidence, even among my own patients and with my own father, that what we're doing works to protect brains. In my father's last five years, my sister Brenda lived with him and was his caretaker. Every two weeks, a doctor would administer his Repatha® injection, and Brenda told me, "It's like the lights went back on." His LDL cholesterol measured at zero—not meaning that he had absolutely no cholesterol in his body, but meaning that it was no longer floating around in his bloodstream with the potential of causing inflammation and forming plaque. Into his late nineties, he was still playing music for audiences, writing children's books, walking in parades, and actively involved with his family. I believe we have Repatha® and the proactive strategies we implemented as part of the Boone Heart Protocol to thank for his long and active life.

I recently saw this with another patient as well. One of my colleagues' wives was suffering from advanced dementia, particularly what's called "sundowner's syndrome," where she became highly agitated every evening. At sundown, it was like his sweet wife turned into Mr. Hyde, with aggressiveness, confusion, and anxiety. This condition is often seen in people with advanced dementia. In addition, we discovered that she had ApoE 4/4 genotype. My colleague came to me for help, and we started her on weekly doses of Repatha®. Suddenly, the kind and intelligent wife he loved was back. It was a startling transformation. Not only that, but she was so inspired by her experience that she wanted to work with the companies that make these drugs (Novartis, Regeneron and Amgen), to tell the world about the power of these medications.

Other patients, an accountant and his wife, are in their seventies. They noticed that they were beginning to experience some borderline cognitive deficits—things we usually consider normal for people their age: memory loss, some confusion, etc. But as soon as we put them on Repatha® and cleaned out the arteries in their brains, they reported feeling sharper than they had in decades. The accountant told me that now, in his seventies, he is doing some of his best work and plans to continue doing the work he loves for a long while yet. These dramatic transformations are unusual, but not uncommon. They tend to occur in people with more significant genetic vascular abnormalities like ApoE 4 and Lp(a). My case is more typical: I've taken PCSK9 inhibitors since 2015 and have not noticed a dramatic brain awakening. On the other hand, I'm not getting any dumber, either.

Gary Ross, the famed director of *The Hunger Games* who also wrote the movie *Big*, is another patient who has seen dramatic results in cognition under our care. He's taking the Multimmunity vitamins and Repatha® and has said that he's working the *New York Times* daily crossword puzzle in half the time it used to take him, cutting his average speed from three hours to just an hour and a half.

I could share dozens more stories of patients who have had similar experiences, almost a "reawakening" as they find themselves able to return to the activities, work, and family life that they love and want to continue enjoying for years to come. They report an immediate, noticeable improvement in cognitive function, especially those who have ApoE4 or LP(a). The punchline of every research study and article is "more research needed." I expect that as research continues and further longitudinal studies are completed, we'll see even further conclusive evidence to help us understand the potential of healing heart disease generally, and certain medications more specifically, to prevent and heal cerebrovascular diseases as well.

I often share an analogy about brain health with my patients. Imagine your brain as a big tree. As you get older, the tree can no longer pull water up from the roots to nourish the outermost branches and leaves. Eventually, the leaves turn brown and the branches begin to wither and die. But if you turn on the sprinklers and make sure the roots have plenty of water, the tree comes back to life and the leaves turn green again. The same thing is happening with our brains. And I've found that with the Boone Heart Protocol, we can make sure that the brain is receiving enough of that life-giving blood to bring it back, maybe even better than before.

Other Inflammatory Diseases

It probably won't surprise you to learn that as we heal the blood vessels through the Boone Heart Protocol and cool the body's overall rates of inflammation, there are innumerable benefits for aiding in recovery from other chronic inflammatory physiological and neurological diseases like multiple sclerosis, Parkinson's disease, Crohn's disease, rheumatoid arthritis, lupus, and many others. Lower LDL cholesterol levels, reduced inflammation, better blood flow, improved heart function, angiogenesis happening as it should—the structure and foundation of the house is sturdy, the doors function as they should, the plumbing is clear, and the electricity is good. The fact is, it's hard for things to go wrong when you're taking care of your heart with the right insights, medications, and lifestyle changes we've seen and recommended for decades.

We have to be careful, though, not to paint the Boone Heart Protocol as a panacea. Of course, every disease has its own pathology and is impacted by the unique circumstances, environment, genetics, and other health conditions of the patient. Every disease needs you to think outside the box and get trusted people to help.

But there's no doubt in my mind, and hopefully not in yours either, that what we recommend here will help most people in most circumstances. And, as the technology and medications continue to improve and become more accessible, we'll see even greater gains in both lifespan and healthspan. The idea of living to a healthy one hundred years isn't a distant possibility, but an immediate reality for all of us, including you, the reader.

The question now is, as the great poet Mary Oliver said, "Tell me, what is it you plan to do with your one wild and precious life?"

CONCLUSION

A World without Heart Disease

Let me transport you to Sardinia, an island in the Mediterranean. The middle of the island is mountainous, with rocky peaks towering over the lush green countryside. Dotted along the hills are squat stone structures dating back to the Bronze Age, called *nuraghi*. Sheep graze amongst the villages, watched over by the vigilant *pastore* who shepherd them. On the coastline, you'll find miles of soft sandy beaches, where the turquoise blue waters are clear enough to see a variety of native fish swimming. Tourists come from all over the world to sunbathe on the beaches or hike amongst the mountains. It's like stepping back in time.

But what's more remarkable than the stunning views is the local culture. Stay long enough, and you'll observe the simple rhythms of life here. Generations of Sardinians have cultivated a habit of walking long distances, eating what they fish or forage every day, spending time with family, friends, and neighbors, taking naps, and preparing traditional, nutritious meals. You might notice that there are an extraordinary number of elderly people living in the villages here. In fact, Sardinia is home to one of the world's largest populations of centenarians, boasting nearly ten times the number of men and women living into their hundreds as the United States.

Sardinia is home to one of the world's five Blue Zones, unique areas in the world where the local population tends to have a much longer lifespan than the rest of the world. Researchers Gianni Pes and Michel Poulain, first took notice of Sardinia because of its remarkably high concentration of male centenarians. In most places of the world, women tend to live, on average, five years longer than men. But here in Sardinia, men are just as likely to live into their hundreds as the women. Pes and Poulain circled Sardinia in blue on a map, giving rise to the term "blue zone." Their interest was taken up by Dan Buettner, a National Geographic journalist, who set out to discover other areas in the world where people tend to live longer. Dan identified four more of these Blue Zones: Okinawa, Japan; Ikaria, Greece; Loma Linda, California; and Nicoya, Costa Rica.

Buettner's findings about lifestyles in the Blue Zones have been well documented in several of his books on the topic. People in the Blue Zones tend to move naturally, feel a strong purpose in their lives, have regular rhythms for rest and destressing, and eat highly nutritious meals with an abundance of fresh seasonal produce, whole grains, local dairy, and limited meat.

What interests me about the Blue Zones, however, is that they can give us a glimpse into a world without heart disease. Buettner writes that in Ikaria, Greece, "They experience 20 percent less cancer, half the rate of heart disease, and almost no dementia." Old age is the number-one killer in these Blue Zones, rather than the chronic diseases that plague the rest of us.

Unfortunately, we can't all live in Ikaria or on Sardinia, but there is much more we can do to live like the people do there. Some of the societal and economic factors that hinder many people across the globe from living to a hundred or beyond, like limited access to healthy, local foods, the prevalence of healthcare deserts in both rural and urban areas, and a healthcare system that rewards treatment rather than prevention, are not as easily solved. Those causes will have to be taken up by lawmakers and politicians. Heart disease and other noncommunicable diseases already make up the majority of deaths not only in the United States, but worldwide. Infectious diseases now only account for about 14 percent of deaths worldwide, a sign of the massive improvements we've already made in public health policy and education in recent decades. The next step will be for our public services to devote further resources and attention to tackling the alarming rise of chronic, noncommunicable diseases.

But it is possible for individuals to achieve many of the same outcomes as those in Blue Zones, despite our circumstances. Remember that one of the Blue Zones is in Loma Linda, California, where a group of Seventh Day Adventists live in an environment more similar to what you and I experience than the distant islands and remote villages of other Blue Zones. The Boone Heart Protocol can help you live like you're in a Blue Zone, no matter where you may be located.

There are certain decisions that we all can and must make to prevent heart disease and lower our chances of cancer, dementia, and other diseases. It starts with making wise choices for our health: getting sufficient sleep, eating nutritious foods that lower inflammation, exercising regularly at moderate levels, drinking plenty of water, taking antioxidant vitamins, avoiding excessive drinking, not smoking, and cultivating mindful practices that help to lower stress. With those principles guiding you, you'll be far more likely to live a long and healthy life.

For most of us, it's not enough, though. The reality of our genetics and environment is that healthy choices are just the first step. The next vital step is becoming an advocate for your health. Ask your doctor for an ultrasound of your carotid arteries and the blood vessels in your legs to assess for plaque and calcium deposits. If available, look at your family history—whatever has killed most of them is likely to come for you, too. You can request a blood panel to test whether you have LP(a), ApoE4, or any of the genotypes that put you at higher risk for CVD, or you can schedule one yourself at any number of testing facilities such as Quest Diagnostics or Labcorp. If scheduling a test yourself, we recommend taking your results to a doctor for help interpreting the results and discussing next steps. If you test positive for those genes, have a family history of heart disease, or have other risk factors, such as diabetes, obesity, high blood pressure, and high LDL cholesterol, then talk to your doctor about whether a statin or PCSK9 inhibitor like Repatha® might be a wise and proactive option for you.

The information and solutions are widely available, if you ask for them.

The one thing you can't do is nothing. For most of us, it's easier to keep living life as we've known it. But that is the best way to ensure that the chronic diseases of the world continue to reign. I sometimes liken the four types of heart disease (atherosclerosis, thrombosis, the deterioration of the muscles and valves, and the electrical issues) to the four horsemen of the apocalypse. But by far the "Secretariat" of them all is plaque. In the end, according to cardiologist James Min, MD, a former Cornell professor and current CEO of Cleerly, "Atherosclerosis is all that really matters." At some point, all four horsemen rear their heads and make themselves known in the presence of plaque. Thrombosis, as you've seen, is a direct result of the accumulation and rupture of hot plaque within the lining of the blood vessels. The muscle rarely fails without significant atherosclerosis. Valves clog, calcify, weaken and malfunction mostly due to atherosclerosis. And the electricity dims and flickers, quickened by the presence of atherosclerosis.

Thankfully, I already see signs that things are beginning to change. While cardiovascular disease is still the world's number-one killer and most urgent public health threat, its mortality rates have been slowly declining since the 1950s. Soon, and especially if you heed the warnings in this book and take proactive measures to protect your health, we'll need to start planning for a new way of life altogether—one in which we may very well expect to live and stay relatively healthy into our nineties and hundreds.

Planning for a Long Future

What would you do if you knew that at seventy-five, eighty, or ninety, you would still be physically strong enough to be active, and mentally sharp enough to continue working, driving, or pursuing various hobbies and interests if you wanted to? Although most of us hope for such a future, the reality is that we don't really expect it. The average age people stop driving is seventy-five, just ten to twelve years after the average retirement age of sixty-five for men and sixty-three for women. Keep in mind that the average life expectancy is seventy-eight for women and seventy-three for men. By eighty-five, if we live that long, over half of us will be living with family, 40 percent will be living alone, and about 8 percent will live in assisted living facilities.

If we first and foremost aggressively attack atherosclerosis, we'll be able to keep all four horses (and many others) in the corral for at least a hundred years. You may well look forward to a much longer, healthier, and more independent life than previously anticipated.

We need to start thinking ahead to how this will impact our finances, careers, and even our relationships. Like my accountant client in Chapter 10, what if you were doing some of your best work into your seventies and eighties? Would you mentor younger generations of leaders and provide counsel and advice for those in your profession? Would you begin writing your first book at age seventy-two, like me, making the safe bet that you'll have plenty of time to see it published and to share it with your colleagues and friends? Most of us want to leave a legacy and make an impact. Without heart disease, you'll have more time.

What if you could help your children and even your grandchildren with raising the next generation? If you spend a lot of time around young children like I do, you know that they require a lot of energy, but there's no reason for that to be a hindrance. Many grandparents in the Blue Zones help to raise their grandchildren, to everybody's benefit.

What if you could foster one of the over six hundred thousand children who pass through the U.S. foster care system every year, or adopt one of the more than sixty-four thousand children whose parents' rights have been terminated and who are waiting for families? What if you had time to volunteer more in your community or contribute to a cause you care about? Imagine if older generations contributed *more* to society as they aged, not less.

If you're living independently and traveling more or pursuing hobbies or volunteer work, then you'll need to plan your finances accordingly. Saving and investing are already challenges for many Americans, but they become a much more urgent priority if you expect to live for a long time and hope to do so comfortably. Perhaps the average age of retirement will rise as the elderly choose to continue working and age-related job discrimination will diminish as everyone realizes the untapped potential of older generations.

How would you and your spouse cultivate a positive relationship and invest in each other if you knew that you'd likely be celebrating your seventy-fifth anniversary together in the not-so-distant future?

The ripple effects of ending heart disease will be wider than we can imagine and yield tremendous benefits for everyone the world over.

The Next Frontier: Orthopedic Issues

This doesn't mean, of course, that as we live longer our bodies won't deteriorate. Without atherosclerosis, our hearts will still need valve replacements or electrical tune-ups as the hardware wears down over time. And as we age, the natural decline in our hormones means that our muscles and bones will grow weaker. The next major challenge for the medical community will no longer be ending the noncommunicable diseases we've discussed in this book, but rather prolonging the life and strength of our musculoskeletal system until we reach the end of our days. We'll have to continue to work hard to ensure that our muscles, tendons, ligaments, joints, teeth, and bones endure as long as possible.

The emphasis for those who plan on living to a healthy one hundred should then be on a handful of key health areas. Taking the right supplements is crucial to ensure our hormones stay balanced. Exercise takes on an even greater role, as high-performance longevity demands regular aerobic and resistance training to strengthen our bones and muscles, increase flexibility, and preserve balance. Workouts like yoga, weight lifting, fast walking, and other activities are accessible exercise options that can be modified as we get older.

"Normal for Age"

One patient of mine was an avid cyclist. He was interested enough in maintaining his longevity and health to meet with me yearly for an annual evaluation. Our testing showed that he had several asymptomatic physiological and biochemical abnormalities that concerned me. Each year, I tried to convince him to take better care of himself for the sake of his heart and brain, and each year, he ignored my advice, saying that his active lifestyle was proof of his excellent health.

Then one morning, as he blazed down the trails above Aspen, the front tire of his mountain bike collided with a small boulder. His bike stopped, but the momentum carried him over the handlebars and right into another rock. He woke up in the emergency room. He had suffered a concussion, but no brain bleed or hematoma. He recovered and soon returned to his life of risky mountain biking.

When he brought me the report from his accident three months later, though, I noticed some abnormal findings from the CT scan that might have easily been missed or glossed over. The report noted "cerebral atrophy" and "microvascular infarctions related to ischemic damage at multiple sites." In other words, my patient's brain was shrinking—and not only that, but he had already suffered from multiple micro-strokes without even realizing it!

Even worse was the radiologist's note at the bottom of the report: "Normal for age."

This small note at the end of the report, almost an afterthought for most people, struck me as a sign of how resigned so many people are—even experienced doctors!—to the decline in brain health in old age. Personally, I refuse to accept this as an unavoidable outcome, for myself, my family, my patients, and for you. I plan on being around for a long while yet to enjoy time with my children and grandkids. I hope you do, too.

Let's be clear. Plaque is not normal. Cerebrovascular degeneration, which accounts for the majority of dementia cases, including vascular dementia and Alzheimer's disease, is *not* normal for age. Atherosclerosis is not normal for age. They are common, true, but they should never be considered normal for anyone. Our brains deteriorate right along with our hearts and the rest of our cardiovascular system for reasons that are entirely preventable. We can and we must do better.

With the WAVi test, we were able to determine that my patient's brain, at just sixty-one, was operating with the speed, power, and reaction time of an eighty-five-year-old. I told him that if he didn't make changes to his health and lifestyle immediately, "In a few years, you'll have Swiss cheese between your ears." Thankfully, this time, he listened. We started him on the Boone Heart Protocol, creating a personalized plan to address his genetic and physiological conditions with medications, antioxidant therapies, lifestyle adjustments, and further testing. I'm proud to say that within two years, his WAVi test showed his brain operating at the performance level of someone just forty-two years old. Against my advice, he continues his high-risk lifestyle zipping down mountains and pursuing the hobby that he loves, but I feel a little better knowing he's protected from the micro-strokes that put him at greater risk.

Whatever kills him, it won't be heart disease.

This should be the new normal. Healing your heart, your brain, and every inch of the sixty-thousand-mile network of blood vessels within your body isn't just a pipe dream; it's a realistic, tangible future well within your grasp. Our medical system has the tools, the medications, and the supplements to keep you healthy. In this book, I've given you the knowledge and the resources to advocate for yourself, and to fight for your right and the right of those you love to live a life free from heart disease. The only question now is: will you do it?

The choice is yours.

APPENDIX

Get Started Healing Your Heart

For those who are interested in taking action today, here's a quick guide to help you get started. While the recommendations we make are fairly simple and straightforward, they are not standard practice, and so it's up to you to advocate for your health and ask for the tests and support that you need. Be advised, too, that some tests and medications may not be covered by insurance, and you may incur additional costs. However, many of these recommendations are becoming more affordable and will continue to drop in price as they become more common.

Step 1: Make Healthy Common-Sense Choices

- Nourish your body. Eat a well-balanced diet of protein, fats, whole grains, locally grown seasonal organic vegetables, and as few processed foods as possible. Drink ten to fifteen cups of water per day. For more personalized nutrition recommendations, work with a registered dietitian nutritionist (RDN) to help you create a customized meal plan that works for your body.
- Exercise moderately. Aim for thirty minutes of cardio three to five times a week, with weight lifting or another strength-training exercise two to three times a week. Rest well.
- Aim to get, at minimum, seven hours of sleep every night.
- Take a daily immunity-boosting vitamin. You can find many excellent options available on the market. Talk to your doctor or a nutritionist if you need support choosing the right vitamin for you.

- Regulate your stress. Turn off screens in the evenings and sleep with screens in a different room. Practice meditation, yoga, or another mindfulness practice. Spend five to ten minutes a day doing deep breathing exercises to lower your heart rate and regulate your blood pressure. Seek counseling to help heal the long-term effects of post-traumatic stress and adverse childhood events.
- Educate yourself on heart disease prevention. Read books about the Blue Zones, anti-inflammatory diets, and heart-healthy practices for living.

There is so much you can do to optimize your health and lower the amount of inflammation throughout your body. But remember: Healthy choices are just the first step. To prevent heart disease, you'll need the support of a good medical provider.

Step 2: Advocate for Your Heart

- At your next physical exam, ask your doctor for an ultrasound of your carotid arteries and the veins in your legs to assess for plaque and calcium deposits. For women, ask your OBGYN to look for evidence of breast arterial calcifications (BACs) during your next mammogram. This is not standard procedure, so be sure to let your doctor know that you are interested in knowing.
- If available, look at your family history—the diseases your parents and grandparents suffered from are the ones you are at most risk for, too. You can also request a blood panel to test whether you might have LP(a), ApoE4, or any of the genotypes that put you at higher risk for CVD, or you can schedule one yourself at any number of testing facilities such as Quest Diagnostics or Labcorp. If scheduling a test yourself, we

recommend taking your results to a doctor for help interpreting the results and discussing next steps.

- If you test positive for those genes, have a family history of heart disease, or have other risk factors, such as diabetes, obesity, high blood pressure, and high LDL cholesterol, talk to your doctor about whether a statin or PCSK9 inhibitor like Repatha® might be a wise and proactive option for you.
- Consider seeking WAVi testing to assess your brain's speed, power, and functionality at a facility near you.

The right next steps for you will depend on your personal health history and your current environment and lifestyle. If you find that your doctor is reluctant to answer your questions, seems uninterested in supporting your journey to preventing heart disease, or doesn't know how, seek out a doctor who will help you find answers and make proactive recommendations for your health. Many functional medicine doctors take an integrated approach to supporting your health, combining cutting-edge technologies with in-depth genetic testing and the right nutritional and supplemental support.

Of course, we'd also be happy to welcome you at the Boone Heart Institute in Denver.

Step 3: Share Your Knowledge

We find that once they see the power of the Boone Heart Protocol, many of our clients want the same benefits for their loved ones. Share this book with your friends and family to help them begin their journey to healing their heart today.

Glossary

Ablation: a common treatment for arrhythmias in which a catheter is guided through the blood vessels to the heart, where electrophysiologists may use either extreme heat or extreme cold to purposefully induce scarring along the electrical route in the heart to disrupt the arrhythmias and reroute the electrical signal to its proper pathways from the SA node to the AV node

Adverse Childhood Events (ACEs): a phrase coined by Dr. Vincent Felitti to describe childhood events that may have a negative impact on a child's development and future adult life. ACES include psychological, physical, or emotional abuse; violence against a child's mother; and living with those who were substance abusers, mentally ill or suicidal, or ever imprisoned.

Alzheimer's disease: a common cerebrovascular disease characterized by the formation of amyloid-ß (Aß) plaques in the brain

Angiogenesis: the body's process for creating new blood vessels from endothelial cells (sprouting angiogenesis), or splitting an existing blood vessel to create new pathways (splitting angiogenesis)

Angiogram/Angiography: a type of imaging technology used to identify blockages through arteries or veins

Angioplasty: a surgical procedure used to improve blood flow when a blockage in a blood vessel has been identified. During the procedure, a catheter is inserted into the blood vessels and guided to the site of a blockage. There, a small balloon at the end of the catheter is inflated to press the plaque down and widen the artery.

Aorta: the artery that supplies oxygenated blood from the left ventricle to the rest of the body

Apolipoprotein E: the gene that controls how the body manufactures lipoproteins. Mutations of this gene are commonly associated with higher risk of heart disease and dementia.

Arrhythmia: an irregular heartbeat

Atherosclerosis/Atherosclerotic heart disease: heart disease characterized by the buildup of plaques throughout the cardiovascular system

Atrial fibrillation (AFib): a condition in which the atria seem to quiver or vibrate rather than pump cohesively

Atrioventricular (AV) node: a small fibrous area of the heart which beats in response to a signal from the sino-atrial (SA) node and causes the ventricles to contract

Atrium: the top, smaller left and right chambers of the heart, which are responsible for taking blood in and passing blood to the ventricles (lower chambers)

Bradycardia: a condition in which the heart beats more slowly than it should

Calcification: a bodily reaction in which plaque within the blood vessels cools and hardens, creating a thicker layer in the intima media

Calcium score: a measurement of the rate of calcification of plaque

Cardiovascular disease: an umbrella term for any disease of the cardiovascular system, typically encompassing atherosclerosis, thrombosis, electrical issues, and the degeneration of the muscle and valves

Cerebrovascular disease: an umbrella term for any disease caused by impaired blood flow to the brain, usually including Alzheimer's disease and dementia

Cholesterol: an essential type of lipid, or fat, responsible for a number of important functions, such as digestion and maintaining the integrity of our cell structure. Cholesterol is also involved in processing and regulating stress hormones like cortisol and sex hormones like estrogen, progesterone, and testosterone.

> **HDL (high-density lipoprotein):** lipoproteins responsible for absorbing unused cholesterol and transporting it back to the liver to be eliminated from the bloodstream
>
> **LDL (low-density lipoprotein):** lipoproteins responsible for delivering cholesterol to the cells for use

Circadian rhythm: the body's natural twenty-four-hour cycle that regulates the function of every system in the body

Cortisol: a stress hormone produced by the adrenal glands, responsible for regulating functions in times of stress

Deep vein thrombosis: a condition characterized by blood clots in the legs

Dementia: an umbrella term for cerebrovascular diseases characterized by cognitive decline and memory loss

Diastolic blood pressure: the bottom number in a blood pressure reading, which measures the blood pressure when the heart is at rest between squeezes

Endothelium: the inner lining of the blood vessel, surrounding the lumen

Electrocardiogram (EKG): a test which records the electrical signals of the heart to check for arrhythmias

Factor II prothrombin mutation: the level of prothrombin is found to be elevated in the blood

Factor V Leiden: the factor responsible for shutting off the formation of a clot once the platelets have the damage to the blood vessel under control; a mutation with this gene (Factor V Leiden) means that the clot will continue forming beyond what's necessary

Fibrin: a protein responsible for creating clots and regulating the healing of clots

Fibrinogen: the pre-clot form of fibrin

Foam cells: an inflammatory buildup of macrophages that creates plaque in the blood vessels

Glucagon-like peptide-1 (GLP-1) agonists: a category of drugs which reduce blood sugar by stimulating the production of insulin and decreasing appetite

Hypercoagulability: an elevated tendency of blood to form clots

Hypertrophy: the tendency of a muscle to thicken with use or exercise

Hypertrophic cardiomyopathy: a condition in which the walls of the heart's left ventricle are unnaturally thickened

Inflammation: a normal part of your body's response to injuries and invaders (like germs), which promotes healing. However, inflammation occurring inside the body when there's no injury or invader can cause harm and a range of chronic diseases, including cardiovascular disease.

Intima media: the wall of the blood vessel

Ischemia/Ischemic heart disease: heart disease characterized by the narrowing of the blood vessels, most commonly caused by atherosclerosis

***Karoshi*:** Japanese term for work-related deaths

Limbic system: the oldest part of the brain, responsible for regulating our emotions and instinctual drives

Lipoprotein-a (Lp(a)): pronounced "L-P-little a," a genetic variant in lipoprotein that causes the accumulation of LDL cholesterol and is a risk factor for atherosclerosis

Lumen: the opening within the blood vessel, through which blood passes

Lusitropic effect: the tendency of the heart to stiffen and become less flexible and more rigid

Macrophages: white blood cells which ignite an inflammatory response to kill the LDL particles that get trapped in the intima media of the blood vessel

Major Adverse Cardiovascular Events (MACEs): sudden cardiovascular conditions, such as a heart attack (myocardial infarction),

Mammography: a breast ultrasound, typically used to screen for breast cancer

Myocardial infarction: a condition caused by the sudden rupture of an artery, in which a clot forms and blood flow to the heart is blocked, commonly called a "heart attack"

Myogenic theory: the theory that the heartbeat originates from the heart muscle itself

Neurogenic theory: the theory that the heartbeat originates from a signal in the brain and is passed through the nervous system to the heart

Neuropeptides: signaling neurons in the brain

NT-proBNP: a test designed to predict heart failure

Oxidation: the decomposition of matter, triggered by exposure to oxygen

PCSK9 inhibitors: a category of drugs designed to block the function of PCSK9, a gene that regulates the production of LDL-receptors

Platelets: tiny particles in the blood that coalesce to form a clot

Pulmonary embolisms: a condition in which a blood clot blocks the flow of blood to the lungs, similar to a myocardial infarction

Sino-atrial (SA) node: a small fibrous area of the heart from which the heart's electrical pulse emanates. The SA node is our natural pacemaker, which triggers the atria to contract.

Statins: a category of drug commonly used to treat and prevent heart disease by reducing the number of LDL particles in the bloodstream

Stent: a hollow mesh tube inserted surgically into a blood vessel to help it retain its shape and structure and prevent blockages

Systolic blood pressure: the top number in a blood pressure reading, which measures the strength of the left ventricle squeezing blood out into the aorta

Tachycardia: an abnormally rapid heartbeat

Takotsubo cardiomyopathy: a condition caused by extreme stress, in which the left ventricle balloons out, triggering cardiac arrest

Thrombosis: heart disease characterized by the formation of blood clots throughout the cardiovascular system, most commonly caused by atherosclerosis

Thrombus: blood clot

Venous thromboembolism: a condition where blood clots form in veins

Ventricles: lower, larger chambers of the heart, which are responsible for pumping blood either out to the lungs (right ventricle) or out to the rest of the body (left ventricle)

Ventricular fibrillation: a serious form of tachycardia in which the left ventricle speeds up to a quivering pace, often triggering a sudden cardiac arrest

WAVi Testing: a quantitative electroencephalogram (qEEG) brain scan designed to test the brain's speed, power, and reaction time

Notes

Introduction

"Life Expectancy in the U.S. Dropped for the Second Year in a Row in 2021." Centers for Disease Control and Prevention, August 31, 2022. https://www.cdc.gov/nchs/pressroom/nchs_press_releases/2022/20220831.htm#:~:text=Life%20expectancy%20at%20birth%20for,2020%20to%2073.2%20in%202021.

"Breast Cancer Statistics: How Common Is Breast Cancer?" Breast Cancer Statistics | How Common Is Breast Cancer? | American Cancer Society. Accessed August 23, 2023. https://www.cancer.org/cancer/types/breast-cancer/about/how-common-is-breast-cancer.html#:~:text=Trends%20in%20breast%20cancer%20deaths,-Breast%20cancer%20is&text=The%20chance%20that%20a%20woman,decline%20of%2043%25%20through%202020.

Maqsood MH, Talha KM, Minhas AMK, Fudim M, Khan SS, Butler J, Khan MS. CDC-WONDER Database Analysis of COVID-19 and Cardiovascular Disease-Related Mortality. J Am Coll Cardiol. 2023 May 2;81(17):1743-1745. doi: 10.1016/j.jacc.2023.02.044. PMID: 37100492; PMCID: PMC10124577.

"Health Topics - Heart Disease and Heart Attacks." Centers for Disease Control and Prevention, August 17, 2021. https://www.cdc.gov/policy/polaris/healthtopics/heartdisease/index.html#:~:text=Economic%20Burden,%2C%20medications%2C%20and%20premature%20death.

"Guide to Coronary Artery Bypass Grafting (CABG)." About Coronary Artery Bypass Grafting (CABG) | Columbia Surgery. Accessed August 23, 2023. https://columbiasurgery.org/conditions-and-treatments/coronary-artery-bypass-grafting#:~:text=By%20restoring%20blood%20flow%20to,rate%20of%20almost%2098%20percent.

"NFL Passing Touchdown % Career Leaders." Pro Football Reference. Accessed August 21, 2024. https://www.pro-football-reference.com/leaders/pass_td_perc_career.htm.

Christy L. Avery, Annie Green Howard, and Hazel B. Nichols, "Trends in US Cancer and Heart Disease Mortality, 1999–2018," *Circulation*, accessed August 20, 2023, https://www.ahajournals.org/doi/10.1161/CIRCULATIONAHA.120.051451.

"The Facts About Women and Heart Disease." www.goredforwomen.org. Accessed August 20, 2023. https://www.goredforwomen.org/en/about-heart-disease-in-women/facts.

"Breast Cancer Statistics: How Common Is Breast Cancer?" Breast Cancer Statistics | How Common Is Breast Cancer? | American Cancer Society. Accessed August 23, 2023. https://www.cancer.org/cancer/types/breast-cancer/about/how-common-is-breast-cancer.html#:~:text=Trends%20in%20breast%20cancer%20deaths,-Breast%20cancer%20is&text=The%20chance%20that%20a%20woman,decline%20of%2043%25%20through%202020.

Maqsood MH, Talha KM, Minhas AMK, Fudim M, Khan SS, Butler J, Khan MS. CDC-WONDER Database Analysis of COVID-19 and Cardiovascular Disease-Related Mortality. J Am Coll Cardiol. 2023 May 2;81(17):1743-1745. doi: 10.1016/j.jacc.2023.02.044. PMID: 37100492; PMCID: PMC10124577.

Part I: The Silent Killer

Chapter 1: The History of Heart Disease

Allam A, Thompson R, Wann L, et al. Atherosclerosis in Ancient Egyptian Mummies. J Am Coll Cardiol Img. 2011 Apr, 4 (4) 315–327. https://doi.org/10.1016/j.jcmg.2011.02.002

Mummert A, Esche E, Robinson J, Armelagos GJ. Stature and robusticity during the agricultural transition: evidence from the bioarchaeological record. Econ Hum Biol. 2011 Jul;9(3):284-301. doi: 10.1016/j.ehb.2011.03.004. Epub 2011 Apr 1. PMID: 21507735.

Glatter KA, Finkelman P. History of the Plague: An Ancient Pandemic for the Age of COVID-19. Am J Med. 2021 Feb;134(2):176-181. doi: 10.1016/j.amjmed.2020.08.019. Epub 2020 Sep 24. PMID: 32979306; PMCID: PMC7513766.

Rogers, H.L. Heart disease and the stress hypothesis in the mid-twentieth century: a historical review. *Psicol. Refl. Crít.* **29**, 47 (2016). https://doi.org/10.1186/s41155-016-0053-5

Taylor, C.Barr & Kraemer, Helena & Bragg, Deborah & Miles, Laughton & Rule, Bruce & Savin, W. Michael & DeBusk, Robert. (1982). A new system for long-term recording and processing of heart rate and physical activity in outpatients. Computers and Biomedical Research. 15. 7-17. 10.1016/0010-4809(82)90048-9.

Javaid Iqbal and others, Coronary stents: historical development, current status and future directions, *British Medical Bulletin*, Volume 106, Issue 1, June 2013, Pages 193–211, https://doi.org/10.1093/bmb/ldt009

Chhabra L, Zain MA, Siddiqui WJ. Angioplasty. 2023 Jan 28. In: StatPearls [Internet]. Treasure Island (FL): StatPearls Publishing; 2023 Jan–. PMID: 29763069.

Centers for Disease Control and Prevention (CDC). Decline in deaths from heart disease and stroke—United States, 1900-1999. MMWR Morb Mortal Wkly Rep. 1999 Aug 6;48(30):649-56. PMID: 10488780.

Mahmood SS, Levy D, Vasan RS, Wang TJ. The Framingham Heart Study and the epidemiology of cardiovascular disease: a historical perspective. Lancet. 2014 Mar 15;383(9921):999-1008. doi: 10.1016/S0140-6736(13)61752-3. Epub 2013 Sep 29. PMID: 24084292; PMCID: PMC4159698.

Bruenn HG. Clinical Notes on the Illness and Death of President Franklin D. Roosevelt. *Annals of Internal Medicine.* 1970;72:579–91. Quoted in Mahmood SS, Levy D, Vasan RS, Wang TJ. The Framingham Heart Study and the epidemiology of cardiovascular disease: a historical perspective. Lancet. 2014 Mar 15;383(9921):999-1008. doi: 10.1016/S0140-

6736(13)61752-3. Epub 2013 Sep 29. PMID: 24084292; PMCID: PMC4159698.

Oppenheimer GM. Becoming the Framingham Study 1947-1950. Am J Public Health. 2005 Apr;95(4):602-10. doi: 10.2105/AJPH.2003.026419. PMID: 15798116; PMCID: PMC1449227.

"Our History - Our Story." Centers for Disease Control and Prevention, April 19, 2023. https://www.cdc.gov/museum/history/our-story.html.

"The Heart Study Heard Around the World: Origins of Framingham and Its Continued Legacy." American College of Cardiology, August 19, 2014. https://www.acc.org/latest-in-cardiology/articles/2014/08/19/09/23/the-heart-study-heard-around-the-world-origins-of-framingham-and-its-continued-legacy.

Rogers, H.L. Heart disease and the stress hypothesis in the mid-twentieth century: a historical review. *Psicol. Refl. Crít.* **29**, 47 (2016). https://doi.org/10.1186/s41155-016-0053-5

Gower, Timothy. "The Heart of a President." Proto Magazine, May 12, 2023. https://protomag.com/cardiology/the-heart-of-a-president/.

Tan SY, Kwock E. Paul Dudley White (1886-1973): Pioneer in modern cardiology. Singapore Med J. 2016 Apr;57(4):215-6. doi: 10.11622/smedj.2016075. PMID: 27075760; PMCID: PMC4853490.

"Collection: World War II." SRITA. Accessed August 25, 2023. https://tobacco.stanford.edu/cigarettes/war-aviation/world-war-ii/.

Lasby, Clarence (1997). Eisenhower's heart attack: how Ike beat heart disease and held on to the presidency. Lawrence: University Press of Kansas. p. 124. ISBN 0700608222. OCLC 35814825.

Brody, Jane E. "Heart Attacks: Turmoil Beneath the Calm." *The New York Times*, June 21, 1983. https://www.nytimes.com/1983/06/21/science/heart-attacks-turmoil-beneath-the-calm.html.

Chapter 2: Killing Us Softly: Atherosclerosis

Peter Attia, *Outlive: The Science & Art of Longevity* (New York: Harmony Books, 2023), pg. XX.

Siddartha Mukherjee, *The Song of the Cell: An Exploration of Medicine and the New Human* (New York: Scribner, 2022), pg. XX.

Konstantinov IE, Mejevoi N, Anichkov NM. Nikolai N. Anichkov and his theory of atherosclerosis. Tex Heart Inst J. 2006;33(4):417-23. PMID: 17215962; PMCID: PMC1764970.

Friedman and Friedland, *Medicine's 10 Greatest Discoveries* (New Haven, CT: Yale University Press, 1998), pg. XX.

Warboys CM, Amini N, de Luca A, Evans PC. The role of blood flow in determining the sites of atherosclerotic plaques. F1000 Med Rep. 2011;3:5. doi: 10.3410/M3-5. Epub 2011 Mar 1. PMID: 21654925; PMCID: PMC3096883.

Glagov S, Weisenberg E, Zarins CK, Stankunavicius R, Kolettis GJ. Compensatory enlargement of human atherosclerotic coronary arteries. N Engl J Med. 1987 May 28;316(22):1371-5. doi: 10.1056/NEJM198705283162204. PMID: 3574413.

Boone, J. "Atherosclerosis Imaging in Primary Preventive Cardiology: The Eradication of Heart Attack and Stroke."

Parthasarathy S, Raghavamenon A, Garelnabi MO, Santanam N. Oxidized low-density lipoprotein. Methods Mol Biol. 2010;610:403-17. doi: 10.1007/978-1-60327-029-8_24. PMID: 20013192; PMCID: PMC3315351.

Burke Harris, Nadine. *The Deepest Well: Healing the Long-Term Effects of Childhood Trauma and Adversity.*

Nobles, Charlie. "Autopsy Shows Chiefs' Thomas Died of a Blood Clot." *The New York Times*, February 10, 2000. https://www.nytimes.com/2000/02/10/sports/pro-football-autopsy-shows-chiefs-thomas-died-of-a-blood-clot.html.

Gross, Jane. "James F. Fixx Dies Jogging: Author on Running Was 52." *The New York Times*, July 22, 1984.

https://www.nytimes.com/1984/07/22/obituaries/james-f-fixx-dies-jogging-author-on-running-was-52.html.

Restak, Richard M. "The Jim Fixx Neurosis: Running Yourself to Death." *The Washington Post*. Accessed August 30, 2023. https://www.washingtonpost.com/archive/opinions/1984/07/29/the-jim-fixx-neurosis-running-yourself-to-death/681bd977-8295-4d4a-802c-bbfd54684be5/.

Behbodikhah J, Ahmed S, Elyasi A, Kasselman LJ, De Leon J, Glass AD, Reiss AB. Apolipoprotein B and Cardiovascular Disease: Biomarker and Potential Therapeutic Target. Metabolites. 2021 Oct 8;11(10):690. doi: 10.3390/metabo11100690. PMID: 34677405; PMCID: PMC8540246.

Witztum JL, Ginsberg HN. Lipoprotein (a): Coming of Age at Last. J Lipid Res. 2016 Mar;57(3):336-9. doi: 10.1194/jlr.E066985. Epub 2016 Feb 2. PMID: 26839334; PMCID: PMC4766984.

"The Nobel Prize in Physiology or Medicine 1985: Press Release." NobelPrize.org. Accessed August 31, 2023. https://www.nobelprize.org/prizes/medicine/1985/press-release/.

Riyaz S. Patel, Stefano Masi, Stefano Taddei, Understanding the role of genetics in hypertension, *European Heart Journal*, Volume 38, Issue 29, 01 August 2017, Pages 2309–2312, https://doi.org/10.1093/eurheartj/ehx273

Huang J, Yang T, Gulliver J, et al. Road Traffic Noise and Incidence of Primary Hypertension. JACC Adv. 2023 Mar, 2 (2) . https://doi.org/10.1016/j.jacadv.2023.100262

Yuan G, Wang J, Hegele RA. Heterozygous familial hypercholesterolemia: an underrecognized cause of early cardiovascular disease. CMAJ. 2006 Apr 11;174(8):1124-9. doi: 10.1503/cmaj.051313. PMID: 16606962; PMCID: PMC1421462.

Chapter 3: The Tipping Point: Thrombosis

Siddartha Mukherjee, *The Song of the Cell: An Exploration of Medicine and the New Human* (New York: Scribner, 2022), pg. XX.

Nadine Burke Harris, *The Deepest Well: Healing the Long-Term Effects of Childhood Trauma and Adversity* (New York: HarperCollins Publishers, 2018), pg. XX.

Brantley, Steve, and Bobbie Myers. "Mount St. Helens – From the 1980 Eruption to 2000." U.S. Geological Survey. Accessed August 31, 2023. https://pubs.usgs.gov/fs/2000/fs036-00/.

Rudolf Virchow, *Cellular Pathology: As Based Upon Physiological and Pathological Histology* (London: John Churchill, 1860), pg. XX.

Félétou, Michel. "Chapter 2: Multiple Functions of the Endothelial Cells." The Endothelium: Part 1: Multiple Functions of the Endothelial Cells-Focus on Endothelium-Derived Vasoactive Mediators., January 1, 1970. https://www.ncbi.nlm.nih.gov/books/NBK57148/#s2.3.

Rumbaut, Rolando E. "Platelet-Vessel Wall Interactions in Hemostasis and Thrombosis." National Center for Biotechnology Information, January 1, 1970. https://www.ncbi.nlm.nih.gov/books/NBK53450/.

Litvinov RI, Weisel JW. What Is the Biological and Clinical Relevance of Fibrin? Semin Thromb Hemost. 2016 Jun;42(4):333-43. doi: 10.1055/s-0036-1571342. Epub 2016 Apr 7. PMID: 27056152; PMCID: PMC5536100.

Caine GJ, Stonelake PS, Lip GY, Kehoe ST. The hypercoagulable state of malignancy: pathogenesis and current debate. Neoplasia. 2002 Nov-Dec;4(6):465-73. doi: 10.1038/sj.neo.7900263. PMID: 12407439; PMCID: PMC1550339.

Sucker C, Geisen C, Schmitt U, Zawislak B. Hypofibrinogenemia and miscarriage: report of a first successful pregnancy under fibrinogen substitution and short review of the literature. Arch Clin Cases. 2022 Sep 26;9(3):100-103. doi: 10.22551/2022.36.0903.10211. PMID: 36176499; PMCID: PMC9512129.

James, Andra H. "Pregnancy-Associated Thrombosis." Hematology, ASH Education Program, January 1, 2009. https://ashpublications.org/hematology/article/2009/1/277/19806/Pregnancy-associated-thrombosis.

Toribio-Fernández, Raquel, Catarina Tristão-Pereira, Juan Carlos Silla-Castro, Sergio Callejas, Belen Oliva, Irene Fernandez-Nueda, Ines Garcia-Lunar, et al. "Apolipoprotein E-E2 and Resistance to Atherosclerosis in Midlife: The Pesa Observational Study." *Circulation Research* 134, no. 4 (January 23, 2024): 411–24. https://doi.org/10.1161/circresaha.123.323921.

Breznican, Anthony. "Chris Hemsworth Changed His Life after an Ominous Health Warning." *Vanity Fair*, November 17, 2022. https://www.vanityfair.com/hollywood/2022/11/chris-hemsworth-exclusive-interview-alzheimers-limitless.

Ornstein, Deborah L., and Mary Cushman. "Factor v Leiden." *Circulation* 107, no. 15 (April 22, 2003). https://doi.org/10.1161/01.cir.0000068167.08920.f1.

Albagoush, Sara A. "Factor v Leiden Mutation." StatPearls [Internet]., April 8, 2023. https://www.ncbi.nlm.nih.gov/books/NBK534802/#:~:text=Heterozygosity%20of%20the%20factor%20V,roughly%2010%25%20to%2020%25.

"Factor II Prothrombin Mutation." Anatolia Geneworks, June 22, 2023. https://www.anatoliageneworks.com/diseases/factor-ii-mutation/.

Tsao, Connie W., Aaron W. Aday, Zaid I. Almarzooq, Cheryl A.M. Anderson, Pankaj Arora, Christy L. Avery, Carissa M. Baker-Smith, et al. "Heart Disease and Stroke Statistics—2023 Update: A Report from the American Heart Association." *Circulation* 147, no. 8 (February 21, 2023). https://doi.org/10.1161/cir.0000000000001123.

"CPR Facts and Stats." cpr.heart.org. Accessed September 11, 2023. https://cpr.heart.org/en/resources/cpr-facts-and-stats/.

Blewer AL, Ibrahim SA, Leary M, Dutwin D, McNally B, Anderson ML, Morrison LJ, Aufderheide TP, Daya M, Idris AH, Callaway CW, Kudenchuk PJ, Vilke GM, Abella BS. Cardiopulmonary Resuscitation Training Disparities in the United States. J Am Heart Assoc. 2017 May 17;6(5):e006124. doi: 10.1161/JAHA.117.006124. PMID: 28515114; PMCID: PMC5524119.

Brown, Lorrel E., and Henry Halperin. "CPR Training in the United States: The Need for a New Gold Standard (and the Gold to Create It)." *Circulation*

Research 123, no. 8 (September 28, 2018): 950–52. https://doi.org/10.1161/circresaha.118.313157.

"Mortality in the United States, 2021." Centers for Disease Control and Prevention, December 8, 2022. https://www.cdc.gov/nchs/products/databriefs/db456.htm.

"Stroke Facts." Centers for Disease Control and Prevention, May 4, 2023. https://www.cdc.gov/stroke/facts.htm#:~:text=The%20death%20rate%20for%20stroke,41.1%20per%20100%2C000%20in%202021.

Pu, Liyuan, Li Wang, Ruijie Zhang, Tian Zhao, Yannan Jiang, and Liyuan Han. "Projected Global Trends in Ischemic Stroke Incidence, Deaths and Disability-Adjusted Life Years From 2020 to 2030." *Stroke* 54, no. 5 (May 2023): 1330–39. https://doi.org/10.1161/strokeaha.122.040073.

Alsaloum, Marissa, Robert S. Zilinyi, Mahesh Madhavan, Daniel J. Snyder, Danial Saleem, John B. Burton, Erika B. Rosenzweig, et al. "Gender Disparities in Presentation, Management, and Outcomes of Acute Pulmonary Embolism." *The American Journal of Cardiology* 202 (July 7, 2023): 67–73. https://doi.org/10.1016/j.amjcard.2023.06.012.

Chapter 4: Wear and Tear: Muscle & Valves

Restak, Richard M. "The Jim Fixx Neurosis: Running Yourself to Death." *The Washington Post*. Accessed August 30, 2023. https://www.washingtonpost.com/archive/opinions/1984/07/29/the-jim-fixx-neurosis-running-yourself-to-death/681bd977-8295-4d4a-802c-bbfd54684be5/.

Reynolds, S. W. A. "The historical struggle for dominance between the heart, liver, and brain". The Proceedings of the 16th Annual History of Medicine Days, March 30th and 31st, 2007 Health Sciences Centre, Calgary, AB. http://hdl.handle.net/1880/47541.

Robinson, Andrew. "Galen: Life Lessons from Gladiatorial Contests." *The Lancet* 382, no. 9904 (November 2013): 1548. https://doi.org/10.1016/s0140-6736(13)62314-4.

Carmona, Pippo. "Who Really Discovered How the Heart Works?" JSTOR Daily, November 20, 2019. https://daily.jstor.org/who-really-discovered-how-the-heart-works/.

Aird WC. Discovery of the cardiovascular system: from Galen to William Harvey. J Thromb Haemost. 2011 Jul;9 Suppl 1:118-29. doi: 10.1111/j.1538-7836.2011.04312.x. PMID: 21781247.

Bolli, Roberto. "William Harvey and the Discovery of the Circulation of the Blood." *Circulation Research* 124, no. 9 (April 25, 2019): 1300–1302. https://doi.org/10.1161/circresaha.119.314977.

Gregory, Andrew. "William Harvey." *Encyclopædia Britannica*, April 22, 2024. https://www.britannica.com/biography/William-Harvey.

Ribatti D. William Harvey and the discovery of the circulation of the blood. J Angiogenes Res. 2009 Sep 21;1:3. doi: 10.1186/2040-2384-1-3. PMID: 19946411; PMCID: PMC2776239.

Friedland G. Discovery of the function of the heart and circulation of blood. Cardiovasc J Afr. 2009 May-Jun;20(3):160. PMID: 19575077; PMCID: PMC3721262.

Butzner M, Leslie DL, Cuffee Y, Hollenbeak CS, Sciamanna C, Abraham T. Stable Rates of Obstructive Hypertrophic Cardiomyopathy in a Contemporary Era. Front Cardiovasc Med. 2022 Jan 6;8:765876. doi: 10.3389/fcvm.2021.765876. PMID: 35071348; PMCID: PMC8770922.

"Patent Foramen Ovale (PFO)." www.heart.org, January 16, 2024. https://www.heart.org/en/health-topics/congenital-heart-defects/about-congenital-heart-defects/patent-foramen-ovale-pfo.

"SADS Conditions." SADS Foundation, May 17, 2022. https://sads.org/sads-conditions/.

"Remember LMU's Hank Gathers 30 Years after His Tragic Death on the Basketball Court | E:60." YouTube, March 4, 2020. https://www.youtube.com/watch?v=Gk-GWuyx3G4.

Physical Activity and Your Heart: Benefits, March 24, 2022. https://www.nhlbi.nih.gov/health/heart/physical-activity/benefits.

Nieman, David C. Exercise, upper respiratory tract infection, and the immune system. Medicine & Science in Sports & Exercise 26(2):p 128-139, February 1994.

"Basketball Legend Kareem Abdul-Jabbar Shares His AFIB Diagnosis Story." stopafib.org, February 24, 2023. https://www.stopafib.org/afib-news-events/news/basketball-legend-kareem-abdul-jabbar-shares-his-afib-diagnosis-story/.

Nath LC, Elliott AD, Weir J, Curl P, Rosanowski SM, Franklin S. Incidence, recurrence, and outcome of postrace atrial fibrillation in Thoroughbred horses. J Vet Intern Med. 2021 Mar;35(2):1111-1120. doi: 10.1111/jvim.16063. Epub 2021 Feb 19. PMID: 33604980; PMCID: PMC7995445.

Chapter 5: When the Lights Go Out: Electrical Failure

Silverman, Mark E, and Charles B Upshaw. "Walter Gaskell and the Understanding of Atrioventricular Conduction and Block." *Journal of the American College of Cardiology* 39, no. 10 (May 14, 2002): 1574–80. https://doi.org/10.1016/s0735-1097(02)01839-9.

Silverman, Mark E., Daniel Grove, and Charles B. Upshaw. "Why Does the Heart Beat?" *Circulation* 113, no. 23 (June 13, 2006): 2775–81. https://doi.org/10.1161/circulationaha.106.616771.

Fye, W B. "The Origin of the Heart Beat: A Tale of Frogs, Jellyfish, and Turtles." *Circulation* 76, no. 3 (September 1987): 493–500. https://doi.org/10.1161/01.cir.76.3.493.

Schweitzer, P. "Jan Evangelista Purkinje (Purkině)." *Clinical Cardiology* 14, no. 1 (January 1991): 85–86. https://doi.org/10.1002/clc.4960140119.

Romanes, George. J. *Jelly-Fish, Star-Fish, and Sea Urchins: Being a Research on Primitive Nervous Systems*. New York: D. APPLETON AND COMPANY, 1898.

Patra C, Zhang X, Brady MF. Physiology, Bundle of His. [Updated 2023 May 1]. In: StatPearls [Internet]. Treasure Island (FL): StatPearls Publishing;

2024 Jan-. Available from: https://www.ncbi.nlm.nih.gov/books/NBK531498/.

"Wolff-Parkinson-White Syndrome." Johns Hopkins Medicine, August 8, 2021. https://www.hopkinsmedicine.org/health/conditions-and-diseases/wolff-parkinson-white-syndrome#:~:text=Wolff%2DParkinson%2DWhite%20syndrome%20(,organized%20way%20control%20your%20heartbeat.

Schnabel, Renate B, Xiaoyan Yin, Philimon Gona, Martin G Larson, Alexa S Beiser, David D McManus, Christopher Newton-Cheh, et al. "50 Year Trends in Atrial Fibrillation Prevalence, Incidence, Risk Factors, and Mortality in the Framingham Heart Study: A Cohort Study." *The Lancet* 386, no. 9989 (May 7, 2015): 154–62. https://doi.org/10.1016/s0140-6736(14)61774-8.

Gillis, Anne M. "Atrial Fibrillation and Ventricular Arrhythmias." *Circulation* 135, no. 6 (February 7, 2017): 593–608. https://doi.org/10.1161/circulationaha.116.025312.

Kornej, Jelena, Christin S. Börschel, Emelia J. Benjamin, and Renate B. Schnabel. "Epidemiology of Atrial Fibrillation in the 21st Century." *Circulation Research* 127, no. 1 (June 18, 2020): 4–20. https://doi.org/10.1161/circresaha.120.316340.

"Long QT Syndrome." National Heart, Lung, and Blood Institute, March 24, 2022. https://www.nhlbi.nih.gov/health/long-qt-syndrome#:~:text=Long%20QT%20syndrome%20(LQTS)%20is,records%20your%20heart%27s%20electrical%20activity.

Ludhwani, Dipesh, Amandeep Goyal, and Mandar Jagtap. "Ventricular Fibrillation." StatPearls [Internet]., August 8, 2023. https://www.ncbi.nlm.nih.gov/books/NBK537120/.

Panagiotou, Demetrios. Famous Hearts and Their Problems. Accessed September 21, 2023. https://www.keystonecardiology.com/blog/famous-hearts-and-their-problems.

Sayed, Moustafa El, Amandeep Goyal, and Avery L. Callahan. "Brugada Syndrome." StatPearls [Internet]., August 8, 2023.

https://www.ncbi.nlm.nih.gov/books/NBK519568/#:~:text=Brugada%20syndrome%20is%20a%20rare,leads%20(V1%2DV3).

Morgan, Emmanuel, and Ken Belson. "Damar Hamlin of Buffalo Bills in Critical Condition After Collapsing During N.F.L. Game." *The New York Times*, January 3, 2023. https://www.nytimes.com/2023/01/02/sports/football/damar-hamlin-bills-hit.html.

Merschel, Michael. "What Is Commotio Cordis, Which NFL Player Damar Hamlin Says Stopped His Heart?" www.heart.org, April 18, 2023. https://www.heart.org/en/news/2023/04/18/what-is-commotio-cordis-which-nfl-player-damar-hamlin-says-stopped-his-heart.

Link, Mark S. "Commotio Cordis." *Circulation: Arrhythmia and Electrophysiology* 5, no. 2 (April 1, 2012): 425–32. https://doi.org/10.1161/circep.111.962712.

"What Is Cardiac Arrest?" National Heart Lung and Blood Institute, May 19, 2022. https://www.nhlbi.nih.gov/health/cardiac-arrest#:~:text=Cardiac%20arrest%20causes%20about%20300%2C000,with%20heart%20attack%20and%20stroke.

Silversmith, David. "Sudden Cardiac Death in Young Athletes - Is It Stoppable?" HealthWorldNet. Accessed September 15, 2023. https://healthworldnet.com/blogs/health-musings/sudden-cardiac-death-in-young-athletes-is-it-stoppable/.

Yow, Allison G., Venkat Rajasurya, and Sandeep Sharma. "Sudden Cardiac Death." StatPearls [Internet]., August 8, 2022. https://www.ncbi.nlm.nih.gov/books/NBK507854/#:~:text=Introduction,to%20maintain%20perfusion%20and%20life.

Hansen, Chris. Marshall Fire Map. Accessed September 17, 2023. https://www.marshallfiremap.com/.

Dougherty, Michael T., and Curtis Johnson. Marshall Fire Investigative Summary and Review, June 8, 2023. https://assets.bouldercounty.gov/wp-content/uploads/2023/06/marshall-fire-investigative-summary.pdf.

Chapter 6: The Fatal Cost of Stress

LeWine, Howard E., ed. "Takotsubo Cardiomyopathy (Broken-Heart Syndrome)." Harvard Health Publishing, June 13, 2023. https://www.health.harvard.edu/heart-health/takotsubo-cardiomyopathy-broken-heart-syndrome.

Boone JL, Anthony JP. Evaluating the impact of stress on systemic disease: the MOST protocol in primary care. J Am Osteopath Assoc. 2003 May;103(5):239-46. PMID: 12776765.

"Heart Disease in the Fire Service." IAFF.org. Accessed February 12, 2024. https://www.hsph.harvard.edu/wp-content/uploads/sites/143/2013/12/IAFF_HeartDiseaseManual_2013.pdf.

Allesøe K, Hundrup YA, Thomsen JF, Osler M. Psychosocial work environment and risk of ischaemic heart disease in women: the Danish Nurse Cohort Study. Occup Environ Med. 2010 May;67(5):318-22. doi: 10.1136/oem.2008.043091. PMID: 20447987.

Theorell T, Jood K, Järvholm LS, Vingård E, Perk J, Östergren PO, Hall C. A systematic review of studies in the contributions of the work environment to ischaemic heart disease development. Eur J Public Health. 2016 Jun;26(3):470-7. doi: 10.1093/eurpub/ckw025. Epub 2016 Mar 31. PMID: 27032996; PMCID: PMC4884330.

Ke, Der-Shin. Overwork, Stroke, and Karoshi-death from Overwork, June 2012. http://www.ant-tnsjournal.com/Mag_Files/21-2/002.pdf.

Muller JE, Ludmer PL, Willich SN, Tofler GH, Aylmer G, Klangos I, Stone PH. Circadian variation in the frequency of sudden cardiac death. Circulation. 1987 Jan;75(1):131-8. doi: 10.1161/01.cir.75.1.131. PMID: 3791599.

Goldberg RJ, Brady P, Muller JE, Chen ZY, de Groot M, Zonneveld P, Dalen JE. Time of onset of symptoms of acute myocardial infarction. Am J Cardiol. 1990 Jul 15;66(2):140-4. doi: 10.1016/0002-9149(90)90577-n. PMID: 2371945.

Morris CJ, Yang JN, Scheer FAJL. The impact of the circadian timing system on cardiovascular and metabolic function. Prog Brain Res. 2012;199:337-

358. doi: 10.1016/B978-0-444-59427-3.00019-8. PMID: 22877674; PMCID: PMC3704149.

"Here's Your Wake-up Call: Daylight Saving Time May Impact Your Heart Health." American Heart Association, March 1, 2024. https://newsroom.heart.org/news/heres-your-wake-up-call-daylight-saving-time-may-impact-your-heart-health#:~:text=While%20gaining%20extra%20daylight%20may,switch%20to%20daylight%20saving%20time.

"Heart Attack Deaths More Likely During Winter Holiday Season Than Any Other Time of Year." American Heart Association, December 7, 2021. https://newsroom.heart.org/news/heart-attack-deaths-more-likely-during-winter-holiday-season-than-any-other-time-of-year.

"Divorce Linked to Higher Risk of Heart Attack." Harvard Health Publishing, June 12, 2015. https://www.health.harvard.edu/heart-health/divorce-linked-to-higher-risk-of-heart-attack.

Felitti VJ, Anda RF, Nordenberg D, Williamson DF, Spitz AM, Edwards V, Koss MP, Marks JS. Relationship of childhood abuse and household dysfunction to many of the leading causes of death in adults. The Adverse Childhood Experiences (ACE) Study. Am J Prev Med. 1998 May;14(4):245-58. doi: 10.1016/s0749-3797(98)00017-8. PMID: 9635069.

Burke Harris, Nadine. *The Deepest Well: Healing the Long-Term Effects of Childhood Trauma and Adversity.* Pg. 38

Churchwell, Keith, Mitchell S.V. Elkind, Regina M. Benjamin, April P. Carson, Edward K. Chang, Willie Lawrence, Andrew Mills, et al. "Call to Action: Structural Racism as a Fundamental Driver of Health Disparities: A Presidential Advisory From the American Heart Association." *Circulation* 142, no. 24 (November 10, 2020): 454–68. https://doi.org/10.1161/cir.0000000000000936.

Van der Kolk, Bessel. *The Body Keeps the Score.* New York: Penguin Books, 2014. Pg. 53.

Su S, Jimenez MP, Roberts CT, Loucks EB. The role of adverse childhood experiences in cardiovascular disease risk: a review with emphasis on plausible mechanisms. Curr Cardiol Rep. 2015 Oct;17(10):88. doi: 10.1007/s11886-015-0645-1. PMID: 26289252; PMCID: PMC4941633.

Münzel T, Gori T, Babisch W, Basner M. Cardiovascular effects of environmental noise exposure. Eur Heart J. 2014 Apr;35(13):829-36. doi: 10.1093/eurheartj/ehu030. Epub 2014 Mar 9. PMID: 24616334; PMCID: PMC3971384.

Lemstra M, Rogers M, Moraros J. Income and heart disease: Neglected risk factor. Can Fam Physician. 2015 Aug;61(8):698-704. PMID: 26836056; PMCID: PMC4541436.

Packham C, Butcher E, Williams M, *et al*Cardiovascular risk profiles and the uptake of the NHS Healthcheck programme in male prisoners in six UK prisons: an observational cross-sectional survey*BMJ Open* 2020;10:e033498. doi: 10.1136/bmjopen-2019-033498.

Yao BC, Meng LB, Hao ML, Zhang YM, Gong T, Guo ZG. Chronic stress: a critical risk factor for atherosclerosis. J Int Med Res. 2019 Apr;47(4):1429-1440. doi: 10.1177/0300060519826820. Epub 2019 Feb 24. PMID: 30799666; PMCID: PMC6460614.

Chapter 7: Why Heart Disease Is Winning

Watson T, Webster MWI, Ormiston JA, Ruygrok PN, Stewart JT. Long and short of optimal stent design. Open Heart. 2017 Oct 30;4(2):e000680. doi: 10.1136/openhrt-2017-000680. PMID: 29118997; PMCID: PMC5663262.

"NIH-Funded Studies Show Stents and Surgery No Better Than Medication, Lifestyle Changes at Reducing Cardiac Events." National Institutes of Health, March 30, 2020. https://www.nih.gov/news-events/news-releases/nih-funded-studies-show-stents-surgery-no-better-medication-lifestyle-changes-reducing-cardiac-events.

George, Andrew T., Kyle A. Clark, and Brahmajee K. Nallamothu. "Stent Cases and the Criminalization of Medical Judgment." *Circulation* 140, no. 25 (December 16, 2019): 2051–53. https://doi.org/10.1161/circulationaha.119.042879.

"Guide to Coronary Artery Bypass Grafting (CABG)." About Coronary Artery Bypass Grafting (CABG). Accessed March 1, 2024. https://columbiasurgery.org/conditions-and-treatments/coronary-artery-bypass-grafting#:~:text=Coronary%20bypass%20operations%20are%20performed,pump%20and%20off%2Dpump%20surgery.

"Breast Cancer Statistics: How Common Is Breast Cancer?" Breast Cancer Statistics | How Common Is Breast Cancer? | American Cancer Society. Accessed March 1, 2024. https://www.cancer.org/cancer/types/breast-cancer/about/how-common-is-breast-cancer.html#:~:text=Trends%20in%20breast%20cancer%20deaths,-Breast%20cancer%20is&text=The%20chance%20that%20a%20woman,decline%20of%2043%25%20through%202020.

"Colorectal Cancer Statistics: How Common Is Colorectal Cancer?" Colorectal Cancer Statistics | How Common Is Colorectal Cancer?, January 29, 2024. https://www.cancer.org/cancer/types/colon-rectal-cancer/about/key-statistics.html.

"Cancer of the Colon and Rectum - Cancer Stat Facts." Surveillance, Epidemiology, and End Results Program. Accessed March 4, 2024. https://seer.cancer.gov/statfacts/html/colorect.html.

Martin, Seth S., Aaron W. Aday, Zaid I. Almarzooq, Cheryl A.M. Anderson, Pankaj Arora, Christy L. Avery, Carissa M. Baker-Smith, et al. "2024 Heart Disease and Stroke Statistics: A Report of US and Global Data from the American Heart Association." *Circulation* 149, no. 8 (February 20, 2024). https://doi.org/10.1161/cir.0000000000001209.

"Heart Disease Facts." Centers for Disease Control and Prevention, May 15, 2023. https://www.cdc.gov/heartdisease/facts.htm#:~:text=One%20person%20dies%20every%2033,United%20States%20from%20cardiovascular%20disease.&text=About%20695%2C000%20people%20in%20the,1%20in%20every%205%20deaths.

Kelli, Heval M., Jeong Hwan Kim, Ayman Samman Tahhan, Chang Liu, Yi-An Ko, Muhammad Hammadah, Samaah Sullivan, et al. "Living in Food Deserts and Adverse Cardiovascular Outcomes in Patients with

Cardiovascular Disease." *Journal of the American Heart Association* 8, no. 4 (February 11, 2019). https://doi.org/10.1161/jaha.118.010694.

Brînzac, Monica G, Ellen Kuhlmann, Gilles Dussault, Marius I Ungureanu, Răzvan M Chereches, and Cătălin O Baba. "Defining Medical Deserts—An International Consensus-Building Exercise." *European Journal of Public Health* 33, no. 5 (July 8, 2023): 785–88. https://doi.org/10.1093/eurpub/ckad107.

Nguyen, Amanda, Jeroen van Meijgaard, Sara Kim, and Tori Marsh. Mapping Healthcare Deserts, September 2021. https://assets.ctfassets.net/4f3rgqwzdznj/1XSl43l40KXMQiJUtl0iIq/ad0070ad4534f9b5776bc2c41091c321/GoodRx_Healthcare_Deserts_White_Paper.pdf.

Centers for Disease Control and Prevention. About underlying cause of death, 1999–2018. https://wonder.cdc.gov/ucd-icd10.html. Accessed August 13, 2020.

Quote in Churchwell, Keith, Mitchell S.V. Elkind, Regina M. Benjamin, April P. Carson, Edward K. Chang, Willie Lawrence, Andrew Mills, et al. "Call to Action: Structural Racism as a Fundamental Driver of Health Disparities: A Presidential Advisory From the American Heart Association." *Circulation* 142, no. 24 (November 10, 2020): 454–68. https://doi.org/10.1161/cir.0000000000000936.

"International Health Cost Comparison Report." Health Care Cost Institute, July 2022. https://healthcostinstitute.org/images/pdfs/international_health_cost_comparison_report_2022.pdf.

Giacomino BD, Cram P, Vaughan-Sarrazin M, Zhou Y, Girotra S. Association of Hospital Prices for Coronary Artery Bypass Grafting With Hospital Quality and Reimbursement. Am J Cardiol. 2016 Apr 1;117(7):1101-6. doi: 10.1016/j.amjcard.2016.01.004. Epub 2016 Jan 14. PMID: 26993975; PMCID: PMC5450906.

Rakshit, Shameek, Matthew Rae, Gary Claxton, Krutika Amin, and Cynthia Cox. "The Burden of Medical Debt in the United States." KFF, February 12, 2024. https://www.kff.org/health-costs/issue-brief/the-burden-of-medical-debt-in-the-united-states/#:~:text=This%20analysis%20of%20government%20data,debt%20of%20more%20than%20%2410%2C000.

Napoli, Nicole. “Medical Bills Financially Burden Almost Half of Cardiovascular Disease Patients.” American College of Cardiology, February 12, 2019. https://www.acc.org/About-ACC/Press-Releases/2019/02/12/10/14/Medical-Bills-Financially-Burden-Almost-Half-of-Cardiovascular-Disease-Patients.

Part II: How to Eradicate Heart Disease

Chapter 8: How Diseases Are Cured

NIAID: National Institute of Allergy and Infectious Diseases. “Antiretroviral Drug Discovery and Development,” February 5, 2024. https://www.niaid.nih.gov/diseases-conditions/antiretroviral-drug-development.

amfAR, The Foundation for AIDS Research. “Curing HIV—How Far Have We Come? - amfAR, the Foundation for AIDS Research,” March 25, 2024. https://www.amfar.org/news/how-many-have-been-cured/.

UNAIDS. “Global HIV & AIDS Statistics — Fact Sheet,” n.d. https://www.unaids.org/en/resources/fact-sheet#:~:text=29.8%20million%20people%20were%20accessing,the%20start%20of%20the%20epidemic.

World Health Organization: WHO. “Cardiovascular Diseases,” June 11, 2019. https://www.who.int/health-topics/cardiovascular-diseases#tab=tab_1.

———. “The Top 10 Causes of Death,” December 9, 2020. https://www.who.int/news-room/fact-sheets/detail/the-top-10-causes-of-death.

Roser, Max, Sophie Ochmann, Hannah Behrens, Hannah Ritchie, and Bernadeta Dadonaite. “Eradication of Diseases.” Our World in Data, February 29, 2024. https://ourworldindata.org/eradication-of-diseases.

Luna F, Luyckx VA. Why have Non-communicable Diseases been Left Behind? Asian Bioeth Rev. 2020 Mar 20;12(1):5-25. doi: 10.1007/s41649-020-00112-8. PMID: 33717328; PMCID: PMC7747415.

Thomas M. Daniel, The history of tuberculosis, Respiratory Medicine, Volume 100, Issue 11, 2006, Pages 1862-1870, ISSN 0954-6111, https://doi.org/10.1016/j.rmed.2006.08.006.

Centers for Disease Control and Prevention. "World TB Day History," October 18, 2023.
https://www.cdc.gov/tb/worldtbday/history.htm#:~:text=Cod%20liver%20oil%2C%20vinegar%20massages,and%20Albert%20Schatz%20developed%20streptomycin.

Roguin A. Rene Theophile Hyacinthe Laënnec (1781-1826): the man behind the stethoscope. Clin Med Res. 2006 Sep;4(3):230-5. doi: 10.3121/cmr.4.3.230. PMID: 17048358; PMCID: PMC1570491.

Segre JA. What does it take to satisfy Koch's postulates two centuries later? Microbial genomics and Propionibacteria acnes. J Invest Dermatol. 2013 Sep;133(9):2141-2. doi: 10.1038/jid.2013.260. Epub 2013 Jul 11. PMID: 23842116; PMCID: PMC3775492.

"The World's Sanatorium · Chasing the Cure · University Libraries Online Exhibits," n.d.
https://exhibits.library.du.edu/librariespresents/exhibits/show/chasing-the-cure/chasing-the-cure/the-world-s-sanatorium.

Mazzarello, Paolo. 2020. "A Physical Cure for Tuberculosis: Carlo Forlanini and the Invention of Therapeutic Pneumothorax" *Applied Sciences* 10, no. 9: 3138. https://doi.org/10.3390/app10093138.

Grant, Gwyn R., Jeffrey A. Lederman, and Robert D. Rrandstetter. "T.G. Heaton, Tuberculosis, and Artificial Pneumothorax Once Again, Back to the Future?" CHEST Journal 112, no. 1 (July 1997).
https://doi.org/10.1378/chest.112.1.7.

"Historical Perspectives Centennial: Koch's Discovery of the Tubercle Bacillus," March 19, 1982.
https://www.cdc.gov/mmwr/preview/mmwrhtml/00000222.htm.

Shulman ST. Clemens von Pirquet: A Remarkable Life and Career. J Pediatric Infect Dis Soc. 2017 Nov 24;6(4):376-379. doi: 10.1093/jpids/piw063. PMID: 27794078.

Hawgood BJ. Albert Calmette (1863-1933) and Camille Guérin (1872-1961): the C and G of BCG vaccine. J Med Biogr. 2007 Aug;15(3):139-46. doi: 10.1258/j.jmb.2007.06-15. PMID: 17641786.

Daniel, Thomas M. "The History of Tuberculosis." *Respiratory Medicine* 100, no. 11 (November 1, 2006): 1862–70. https://doi.org/10.1016/j.rmed.2006.08.006.

Woodruff HB. Selman A. Waksman, winner of the 1952 Nobel Prize for physiology or medicine. Appl Environ Microbiol. 2014 Jan;80(1):2-8. doi: 10.1128/AEM.01143-13. Epub 2013 Oct 25. PMID: 24162573; PMCID: PMC3911012.

World Health Organization: WHO. "The Top 10 Causes of Death," December 9, 2020. https://www.who.int/news-room/fact-sheets/detail/the-top-10-causes-of-death.

Zhang X, Xie Q, Ye Z, Li Y, Che Z, Huang M, Zeng J. Mesenchymal Stem Cells and Tuberculosis: Clinical Challenges and Opportunities. Front Immunol. 2021 Jul 22;12:695278. doi: 10.3389/fimmu.2021.695278. PMID: 34367155; PMCID: PMC8340780.

Chackerian, Bryce, and Alan T. Remaley. "PCSK9 Vaccines: A Promising New Strategy for the Treatment of Hypercholesterolemia?" Journal of Lipid Research 65, no. 3 (March 1, 2024): 100524. https://doi.org/10.1016/j.jlr.2024.100524.

American College of Cardiology. "VERVE-101: CRISPR-Based Gene Editing Therapy Shows Promise in Reducing LDL-C and PCSK9 Levels in Patients With HeFH - American College of Cardiology," November 8, 2023. https://www.acc.org/Latest-in-Cardiology/Articles/2023/11/08/20/14/sun-445pm-heart1-aha-2023.

Chapter 9: The Boone Heart Protocol

Harvard Health. "C-Reactive Protein Test to Screen for Heart Disease: Why Do We Need Another Test?," March 21, 2017. https://www.health.harvard.edu/heart-health/c-reactive-protein-test-to-screen-for-heart-disease.

Azam, Shadi, Rulla M. Tamimi, Michele Drotman, Kemi Babagbemi, Allison D. Levy, and Jessica M. Peña. "Assessing Breast Arterial Calcification in Mammograms and Its Implications for Atherosclerotic Cardiovascular Disease Risk." *Clinical Imaging*, March 1, 2024, 110129. https://doi.org/10.1016/j.clinimag.2024.110129.

Suh JW, Yun B. Breast Arterial Calcification: A Potential Surrogate Marker for Cardiovascular Disease. J Cardiovasc Imaging. 2018 Sep;26(3):125-134. doi: 10.4250/jcvi.2018.26.e20. Epub 2018 Sep 18. PMID: 30310879; PMCID: PMC6160812.

Cohen SN, Friedlander AH, Jolly DA, Date L. Carotid calcification on panoramic radiographs: an important marker for vascular risk. Oral Surg Oral Med Oral Pathol Oral Radiol Endod. 2002 Oct;94(4):510-4. doi: 10.1067/moe.2002.125580. PMID: 12374929.

Devesa, Ana, Borja Ibáñez, Waqas Malick, Elizabeth Tinuoye, Jessica Bustamante, Carlos Peyra, Robert S. Rosenson, Deepak L. Bhatt, Gregg W. Stone, and Valentín Fuster. "Primary Prevention of Subclinical Atherosclerosis in Young Adults." *Journal of the American College of Cardiology* 82, no. 22 (November 1, 2023): 2152–62. https://doi.org/10.1016/j.jacc.2023.09.817.

Amezcua-Castillo E, González-Pacheco H, Sáenz-San Martín A, Méndez-Ocampo P, Gutierrez-Moctezuma I, Massó F, Sierra-Lara D, Springall R, Rodríguez E, Arias-Mendoza A, Amezcua-Guerra LM. C-Reactive Protein: The Quintessential Marker of Systemic Inflammation in Coronary Artery Disease-Advancing toward Precision Medicine. Biomedicines. 2023 Sep 2;11(9):2444. doi: 10.3390/biomedicines11092444. PMID: 37760885; PMCID: PMC10525787.

Chou, Roger, Amy Cantor, Tracy Dana, Jesse Wagner, Azrah Y. Ahmed, Rongwei Fu, and Maros Ferencik. "Statin Use for the Primary Prevention of

Cardiovascular Disease in Adults." *JAMA* 328, no. 8 (August 23, 2022): 754. https://doi.org/10.1001/jama.2022.12138.

Matyori A, Brown CP, Ali A, Sherbeny F. Statins utilization trends and expenditures in the U.S. before and after the implementation of the 2013 ACC/AHA guidelines. Saudi Pharm J. 2023 Jun;31(6):795-800. doi: 10.1016/j.jsps.2023.04.002. Epub 2023 Apr 11. PMID: 37228328; PMCID: PMC10203693.

Haiku. "CTT Collaboration," n.d. https://www.cttcollaboration.org/.

Martin, Seth S., Aaron W. Aday, Zaid Almarzooq, Cheryl A.M. Anderson, Pankaj Arora, Christy L. Avery, Carissa M. Baker-Smith, et al. "2024 Heart Disease and Stroke Statistics: A Report of US and Global Data From the American Heart Association." *Circulation* 149, no. 8 (February 20, 2024). https://doi.org/10.1161/cir.0000000000001209.

Thompson W, Morin L, Jarbøl DE, Andersen JH, Ernst MT, Nielsen JB, Haastrup P, Schmidt M, Pottegård A. Statin Discontinuation and Cardiovascular Events Among Older People in Denmark. JAMA Netw Open. 2021 Dec 1;4(12):e2136802. doi: 10.1001/jamanetworkopen.2021.36802. Erratum in: JAMA Netw Open. 2022 Jan 4;5(1):e220010. PMID: 34854906; PMCID: PMC8640890.

Merz, C. Noel Bairey, Holly Andersen, Emily Sprague, Adam Burns, Mark Keida, Mary Norine Walsh, Phyllis Greenberger, et al. "Knowledge, Attitudes, and Beliefs Regarding Cardiovascular Disease in Women." *Journal of the American College of Cardiology* 70, no. 2 (July 1, 2017): 123–32. https://doi.org/10.1016/j.jacc.2017.05.024.

Hajar R. PCSK 9 Inhibitors: A Short History and a New Era of Lipid-lowering Therapy. Heart Views. 2019 Apr-Jun;20(2):74-75. doi: 10.4103/HEARTVIEWS.HEARTVIEWS_59_19. PMID: 31462965; PMCID: PMC6686613.

Hoshino, Masahiro, Taishi Yonetsu, Eisuke Usui, Yoshihisa Kanaji, Hiroaki Ohya, Yohei Sumino, Masao Yoshizumi, et al. "Clinical Significance of the Presence or Absence of Lipid-Rich Plaque Underneath Intact Fibrous Cap Plaque in Acute Coronary Syndrome." *Journal of the American Heart*

Association. Cardiovascular and Cerebrovascular Disease 8, no. 9 (May 7, 2019). https://doi.org/10.1161/jaha.118.011820.

Santosa, Ailiana, Stefan Franzén, Jonatan Nåtman, Björn Wettermark, Ingela Parmryd, and Fredrik Nyberg. "Protective Effects of Statins on COVID-19 Risk, Severity and Fatal Outcome: A Nationwide Swedish Cohort Study." *Scientific Reports* 12, no. 1 (July 14, 2022). https://doi.org/10.1038/s41598-022-16357-2.

Murto, Mika O., Niklas Simolin, Otso Arponen, Aino Siltari, Miia Artama, Kala Visvanathan, Arja Jukkola, and Teemu J. Murtola. "Statin Use, Cholesterol Level, and Mortality Among Females With Breast Cancer." *JAMA Network Open* 6, no. 11 (November 17, 2023): e2343861. https://doi.org/10.1001/jamanetworkopen.2023.43861.

Nicholls SJ, Kataoka Y, Nissen SE, Prati F, Windecker S, Puri R, Hucko T, Aradi D, Herrman JR, Hermanides RS, Wang B, Wang H, Butters J, Di Giovanni G, Jones S, Pompili G, Psaltis PJ. Effect of Evolocumab on Coronary Plaque Phenotype and Burden in Statin-Treated Patients Following Myocardial Infarction. JACC Cardiovasc Imaging. 2022 Jul;15(7):1308-1321. doi: 10.1016/j.jcmg.2022.03.002. Epub 2022 Mar 16. PMID: 35431172.

Li, Zhiyong, Simon Howarth, Tjun Yip Tang, and Jonathan H. Gillard. "How Critical Is Fibrous Cap Thickness to Carotid Plaque Stability?" *Stroke* 37, no. 5 (May 1, 2006): 1195–99. https://doi.org/10.1161/01.str.0000217331.61083.3b.

Bonaca MP, Nault P, Giugliano RP, Keech AC, Pineda AL, Kanevsky E, Kuder J, Murphy SA, Jukema JW, Lewis BS, Tokgozoglu L, Somaratne R, Sever PS, Pedersen TR, Sabatine MS. Low-Density Lipoprotein Cholesterol Lowering With Evolocumab and Outcomes in Patients With Peripheral Artery Disease: Insights From the FOURIER Trial (Further Cardiovascular Outcomes Research With PCSK9 Inhibition in Subjects With Elevated Risk). Circulation. 2018 Jan 23;137(4):338-350. doi: 10.1161/CIRCULATIONAHA.117.032235. Epub 2017 Nov 13. PMID: 29133605.

ABC. "Watch an Oprah Special: Shame, Blame and the Weight Loss Revolution TV Show - ABC.com," n.d. https://abc.com/shows/an-oprah-special-shame-blame-and-the-weight-loss-revolution.

Lincoff AM, Brown-Frandsen K, Colhoun HM, Deanfield J, Emerson SS, Esbjerg S, Hardt-Lindberg S, Hovingh GK, Kahn SE, Kushner RF, Lingvay I, Oral TK, Michelsen MM, Plutzky J, Tornøe CW, Ryan DH; SELECT Trial Investigators. Semaglutide and Cardiovascular Outcomes in Obesity without Diabetes. N Engl J Med. 2023 Dec 14;389(24):2221-2232. doi: 10.1056/NEJMoa2307563. Epub 2023 Nov 11. PMID: 37952131.

Callahan, Alice. "Is Intermittent Fasting Bad for Your Heart? Here's What We Know." *New York Times*. March 21, 2024. https://www.nytimes.com/2024/03/20/well/eat/intermittent-fasting-study.html

Boone, J. The MAVRIC Trial: Micronutrient Antioxidant Vitamin Reversal of the

Intima Media Thickness of the Carotid Artery. Unpublished manuscript.

Childs E, de Wit H. Regular exercise is associated with emotional resilience to acute stress in healthy adults. Front Physiol. 2014 May 1;5:161. doi: 10.3389/fphys.2014.00161. PMID: 24822048; PMCID: PMC4013452.

www.heart.org. "American Heart Association Recommendations for Physical Activity in Adults and Kids," January 19, 2024. https://www.heart.org/en/healthy-living/fitness/fitness-basics/aha-recs-for-physical-activity-in-adults.

Bentley TGK, D'Andrea-Penna G, Rakic M, Arce N, LaFaille M, Berman R, Cooley K, Sprimont P. Breathing Practices for Stress and Anxiety Reduction: Conceptual Framework of Implementation Guidelines Based on a Systematic Review of the Published Literature. Brain Sci. 2023 Nov 21;13(12):1612. doi: 10.3390/brainsci13121612. PMID: 38137060; PMCID: PMC10741869.

Prathikanti S, Rivera R, Cochran A, Tungol JG, Fayazmanesh N, Weinmann E. Treating major depression with yoga: A prospective, randomized, controlled pilot trial. PLoS One. 2017 Mar 16;12(3):e0173869. doi: 10.1371/journal.pone.0173869. PMID: 28301561; PMCID: PMC5354384.

Chapter 10. Healing from Head to Toe

McDevitt J, Krynetskiy E. Genetic findings in sport-related concussions: potential for individualized medicine? Concussion. 2017 Jan 24;2(1):CNC26. doi: 10.2217/cnc-2016-0020. PMID: 30202567; PMCID: PMC6096436.

Professional, Cleveland Clinic Medical. "Angiogenesis." Cleveland Clinic, n.d. https://my.clevelandclinic.org/health/articles/24206-angiogenesis.

"Cholesterol Numbers and What They Mean." Cleveland Clinic, n.d. https://my.clevelandclinic.org/health/articles/11920-cholesterol-numbers-what-do-they-mean.

Li, William. *Eat to Beat Disease.* Pg. 9

Melloni, Chiara, Gretchen G. Kimmick, Taofik Oyekunle, Leah L. Zullig, Michel G Khouri, Lauren Howard, Michael J. Kelley, and Susan F. Dent. "Abstract 15660: Statin Use Is Associated With Increased Overall Survival in Patients With Colorectal Cancer: Findings From a Cohort of 29,498 United States Veterans." *Circulation* 140, no. Suppl_1 (November 11, 2019). https://doi.org/10.1161/circ.140.

Quagliariello V, Bisceglia I, Berretta M, Iovine M, Canale ML, Maurea C, Giordano V, Paccone A, Inno A, Maurea N. PCSK9 Inhibitors in Cancer Patients Treated with Immune-Checkpoint Inhibitors to Reduce Cardiovascular Events: New Frontiers in Cardioncology. Cancers (Basel). 2023 Feb 22;15(5):1397. doi: 10.3390/cancers15051397. PMID: 36900189; PMCID: PMC10000232.

Liu, X., Bao, X., Hu, M. *et al.* Inhibition of PCSK9 potentiates immune checkpoint therapy for cancer. *Nature* 588, 693–698 (2020). https://doi.org/10.1038/s41586-020-2911-7

https://seer.cancer.gov/statfacts/html/common.html#:~:text=Breast%2C%20lung%20and%20bronchus%2C%20prostate,nearly%2050%25%20of%20all%20death.

Callahan CM, Apostolova LG, Gao S, Risacher SL, Case J, Saykin AJ, Lane KA, Swinford CG, Yoder MC. Novel Markers of Angiogenesis in the Setting of Cognitive Impairment and Dementia. J Alzheimers Dis. 2020;75(3):959-969. doi: 10.3233/JAD-191293. PMID: 32390626; PMCID: PMC8351220.

Mazura, Alexander D, Anke Ohler, Steffen E. Storck, Magdalena Kurtyka, Franka Scharfenberg, Sascha Weggen, Christoph Becker-Pauly, and Claus U. Pietrzik. "PCSK9 Acts as a Key Regulator of Aβ Clearance Across the Blood–brain Barrier." *Cellular and Molecular Life Sciences* 79, no. 4 (March 27, 2022). https://doi.org/10.1007/s00018-022-04237-x.

Chandar V, Gidvani CH, Gupta AK, Wilson CG, Sharma YV. LIPID PROFILE IN NORMAL HEALTHY CHILDREN. Med J Armed Forces India. 1994 Apr;50(2):101-104. doi: 10.1016/S0377-1237(17)31008-0. Epub 2017 Jun 27. PMID: 28769178; PMCID: PMC5529694.

Professional, Cleveland Clinic Medical. "Angiogenesis." Cleveland Clinic, n.d. https://my.clevelandclinic.org/health/articles/24206-angiogenesis.

———. "Cholesterol Numbers and What They Mean." Cleveland Clinic, n.d. https://my.clevelandclinic.org/health/articles/11920-cholesterol-numbers-what-do-they-mean.

Marston, Nicholas, Robert P. Giugliano, Jeong-Gun Park, Andrea Ruzza, Peter Sever, Anthony Keech, and Marc S. Sabatine. "Cardiovascular Benefit of Lowering Low-Density Lipoprotein Cholesterol Below 40 Mg/dL." *Circulation* 144, no. 21 (November 23, 2021): 1732–34. https://doi.org/10.1161/circulationaha.121.056536.

Oliver, Mary. "The Summer Day."

Conclusion: A World without Heart Disease

Giuffrida, Angela. "'If You Talk, You Live Well': The Remote Sardinian Village With Eight Centenarians." *The Guardian*, August 10, 2021.

https://www.theguardian.com/world/2021/aug/08/if-you-talk-you-live-well-the-remote-sardinian-village-famed-for-longevity#:~:text=Sardinia%20has%20been%20identified%20as,33.6%20for%20every%20100%2C000%20inhabitants.

Admin. "History of Blue Zones." Blue Zones, August 30, 2023. https://www.bluezones.com/about/history/.

Baum F, Musolino C, Gesesew HA, Popay J. New Perspective on Why Women Live Longer Than Men: An Exploration of Power, Gender, Social Determinants, and Capitals. Int J Environ Res Public Health. 2021 Jan 14;18(2):661. doi: 10.3390/ijerph18020661. PMID: 33466763; PMCID: PMC7829786.

Buettner D, Skemp S. Blue Zones: Lessons From the World's Longest Lived. Am J Lifestyle Med. 2016 Jul 7;10(5):318-321. doi: 10.1177/1559827616637066. PMID: 30202288; PMCID: PMC6125071.

Dattani, Saloni, Fiona Spooner, Hannah Ritchie, and Max Roser. "Causes of Death." Our World in Data, December 28, 2023. https://ourworldindata.org/causes-of-death#key-insights.

Society of Cardiovascular Computed Tomography. "Sponsored Symposium | Why Atherosclerosis Is All That Really Matters," July 17, 2021. https://www.youtube.com/watch?v=B-m9FzIKk0g.

Mensah GA, Wei GS, Sorlie PD, Fine LJ, Rosenberg Y, Kaufmann PG, Mussolino ME, Hsu LL, Addou E, Engelgau MM, Gordon D. Decline in Cardiovascular Mortality: Possible Causes and Implications. Circ Res. 2017 Jan 20;120(2):366-380. doi: 10.1161/CIRCRESAHA.116.309115. PMID: 28104770; PMCID: PMC5268076.

"Yahoo Is Part of the Yahoo Family of Brands," n.d. https://finance.yahoo.com/news/age-most-americans-retire-know-160816310.html?guccounter=1&guce_referrer=aHR0cHM6Ly93d3cuZ29vZ2xlLmNvbS8&guce_referrer_sig=AQAAAIjRvytjyeZg1IQ5QzlUIVBoHGbCH5Qq3Va3vLgZxIEnmu1RKUf5MbSLUBIrjC0-i3EyimctMuNvbmAdpJiKFQhv70SuAO7gK7Hq8lBimQPLbiPZXkrpvjiJq4HknRrlZGkSqtaWz_pIPyn-6YySu9v28LIaSmiMw6Sc53JArh4L.

USAFacts. "How Many Kids Are in Foster Care in the US?" USAFacts, August 23, 2023. https://usafacts.org/articles/how-many-kids-are-in-foster-care/.

About the Author

Jeffrey L. Boone, MD, MS, has been a pioneer in the prevention of heart disease and stroke for the past three decades. As an active clinician and researcher, Dr. Boone is an internationally recognized expert on the effects of mental stress on the heart. He has authored and co-authored numerous professional articles and books pertaining to Preventive Cardiology, Cardiometabolic Risk, and Stress Medicine.

As an international lecturer, Dr. Boone has spoken to thousands of health care professionals around the world. Dr. Boone's unique clinical approach focuses on aggressive prevention of cardiovascular disease, including the evaluation of the cardiovascular consequences of mental stress, the early clinical use of the latest cardiac imaging techniques and the advanced detection and treatment of heart disease and stroke. In recent years, Dr. Boone has begun to explore the implications of his preventive cardiology work on brain health, leading to promising developments in the treatment of dementia, chronic traumatic encephalopathy, and other brain health issues.

Dr. Boone works as a Consultant in Preventive Cardiology for the Denver Broncos and San Francisco 49ers of the NFL, the Colorado Rockies of Major League Baseball, and the Atlanta Hawks of the NBA. Dr. Boone sat on the NFL's Cardiovascular Committee for a decade, and created the NFL Alumni Association's Cardiovascular Screening Program, an ongoing investigation into the cardiovascular health of former NFL players.

A lifelong athlete, Dr. Boone played basketball and tennis in college, and completed three marathons. His latest passion is playing pickleball with his wife, Colette, in far flung locations around the country and the world. When not travelling for work or play, Dr. Boone's time is taken up by his twelve grandchildren.

www.ingramcontent.com/pod-product-compliance
Lightning Source LLC
LaVergne TN
LVHW011633200926
848484LV00014B/2585/J